Learning NServiceBus Sagas

Discover how to design, build, and test sagas and messaging with NServiceBus

Rich Helton

BIRMINGHAM - MUMBAI

Learning NServiceBus Sagas

Copyright © 2015 Packt Publishing

All rights reserved. No part of this book may be reproduced, stored in a retrieval system, or transmitted in any form or by any means, without the prior written permission of the publisher, except in the case of brief quotations embedded in critical articles or reviews.

Every effort has been made in the preparation of this book to ensure the accuracy of the information presented. However, the information contained in this book is sold without warranty, either express or implied. Neither the author, nor Packt Publishing, and its dealers and distributors will be held liable for any damages caused or alleged to be caused directly or indirectly by this book.

Packt Publishing has endeavored to provide trademark information about all of the companies and products mentioned in this book by the appropriate use of capitals. However, Packt Publishing cannot guarantee the accuracy of this information.

First published: January 2015

Production reference: 1240115

Published by Packt Publishing Ltd.
Livery Place
35 Livery Street
Birmingham B3 2PB, UK.

ISBN 978-1-78217-349-6

www.packtpub.com

Credits

Author
Rich Helton

Reviewers
Neil Bourgeois
Prashant Brall
Mark Huber

Commissioning Editor
Usha Iyer

Acquisition Editor
Kevin Colaco

Content Development Editors
Akshay Nair
Priya Singh

Technical Editor
Edwin Moses

Copy Editors
Sarang Chari
Puja Lalwani
Veena Mukundan

Project Coordinator
Mary Alex

Proofreaders
Samuel Redman Birch
Maria Gould
Bernadette Watkins

Indexer
Monica Ajmera Mehta

Graphics
Valentina D'silva

Production Coordinator
Nilesh R. Mohite

Cover Work
Arvindkumar Gupta
Nilesh R. Mohite

About the Author

Rich Helton, as a principal software engineer, builds and reviews large-scale systems and trains hundreds of developers as well.

Rich has spent over 2 decades in designing and building systems. During this time, he has built, architected, and designed multiple systems, as well as managed many different technical teams. He has built many large-scale enterprise solutions using the most popular C# and Java frameworks and has expertise in the financial, aeronautical, and security domains.

Rich's passion for designing and teaching HTML5, ESBs, ORM's test-driven development, NoSQL, iOS, IoCs, and cloud and iPad development was discovered while training developers and architects. He freely shares some of the slides from these trainings on http://www.slideshare.net/rhelton_1.

> I would like to thank my wife, Johennie, and my daughters, Ashley and Courtney, for their ongoing support.

About the Reviewers

Neil Bourgeois is a software engineer at Pentair Technical Solutions, where he leads the Engineering Software team. Solutions he has architected and implemented include a high-volume metering and billing system for the utilities industry and an industry-leading 3D engineering tool for the industrial heat-tracing field. He applies the discipline of Agile software development to his work and believes that great software comes from great team cultures. He strives to lead his teams to great cultures.

Prashant Brall is a principal consultant and senior software architect/developer who uses Microsoft technologies. He works for Veritec Pty Ltd (www.veritec.com.au) in Canberra, Australia, and has been developing software for the past 18 years. He enjoys writing about his experiences on his blog at https://prashantbrall.wordpress.com.

Prashant has also reviewed the book *Instant AutoMapper*, *Packt Publishing*.

In his leisure time, he enjoys watching movies with his wife and playing musical instruments.

> I would like to thank my fantastic wife, Jhumur, for being my best friend and for all the love and support she has given me. I would also like to thank my parents Mr. Hem Brall and Mrs. Prabha Brall for all the guidance, love, and care they have given me.

Mark Huber is a developer, team manager, and general-purpose tech enthusiast who lives and works in Dallas, Texas. In the last several years, he has focused on specializing in large-scale web platforms in the automotive market. He has spent the majority of his time in the .NET environment, but believes the key to mastering your preferred domain is expanding beyond traditional boundaries to look at how other languages and technologies approach the same problems. He has been an avid researcher of other platforms including Java, Ruby, Python, and others. He is also a strong advocate of interweaving .NET projects with supporting open source technologies not traditionally considered with a .NET platform, such as ElasticSearch, Redis, RabbitMQ, Memcached, and others.

Mark is currently a software development manager working at Dealer.com, a Dealertrack technology solution. He is privileged to work with a team of developers, QA, DevOps, product owners, and many others, who truly live the principles and values of the Agile Manifesto (`http://agilemanifesto.org/`) as well as view their work through the lens of the Software Craftsmanship Manifesto (`http://manifesto.softwarecraftsmanship.org/`).

> I would like to extend a heartfelt thank you to my colleagues, mentors, friends, and most importantly my family, who have been an unending stream of support and growth over the years. Without them, I would not be where I am today.

"Perfecting oneself is as much unlearning as it is learning."

- Edsgar Dijkstra

www.PacktPub.com

Support files, eBooks, discount offers, and more

For support files and downloads related to your book, please visit `www.PacktPub.com`.

Did you know that Packt offers eBook versions of every book published, with PDF and ePub files available? You can upgrade to the eBook version at `www.PacktPub.com` and as a print book customer, you are entitled to a discount on the eBook copy. Get in touch with us at `service@packtpub.com` for more details.

At `www.PacktPub.com`, you can also read a collection of free technical articles, sign up for a range of free newsletters and receive exclusive discounts and offers on Packt books and eBooks.

`https://www2.packtpub.com/books/subscription/packtlib`

Do you need instant solutions to your IT questions? PacktLib is Packt's online digital book library. Here, you can search, access, and read Packt's entire library of books.

Why subscribe?

- Fully searchable across every book published by Packt
- Copy and paste, print, and bookmark content
- On demand and accessible via a web browser

Free access for Packt account holders

If you have an account with Packt at `www.PacktPub.com`, you can use this to access PacktLib today and view 9 entirely free books. Simply use your login credentials for immediate access.

Table of Contents

Preface	**1**
Chapter 1: Introduction to Sagas	**9**
A brief introduction to ESBs	**10**
Event-driven jobs	14
Additional SOA patterns	15
The publish-subscribe pattern	15
The request-reply pattern	17
The gateway pattern	18
The source code	19
The DataBus pattern	21
Timeout patterns	22
Message mutation patterns	28
The source code	29
Message encryption patterns	30
The source code	30
The ScaleOut pattern	32
The saga design pattern	32
Sagas – what are they good for?	**34**
Summary	**37**
Chapter 2: NServiceBus Saga Architecture	**39**
Upgrading from NSB version 4 to 5	**39**
The saga workflow	**41**
Message flow	48
Deployment	**55**
ServiceInsight	**56**
Summary	**58**

Table of Contents

Chapter 3: The Particular Service Platform — 59
Introducing NSB components — 59
Understanding ServicePulse and its function — 61
Understanding ServiceControl and its function — 62
Understanding ServiceInsight and its function — 66
Creating a ServiceMatrix solution — 69
Sagas through ServiceMatrix — 84
Introducing CustomChecks for ServicePulse — 88
Summary — 91

Chapter 4: Saga Development — 93
A brief overview of ASP.NET MVC — 94
Sagas and web services — 96
The source code — 96
Creating a WCF server — 97
Adding messages — 100
Adding the message handler — 102
Adding the configuration — 103
Adding tracing — 106
Viewing the web service — 110
Considerations when deploying — 110
Creating a WCF client — 111
Adding the service reference — 112
Calling the service reference — 114
Revisiting the design — 116
The source code — 118
Adding NServiceBus to MVC — 120
Message handler unit testing — 123
Saga handler unit testing — 128
Integration tests with MVC — 130
RabbitMQ for NSB — 135
The source code — 137
Changing the endpoints — 137
ActiveMQ in NSB — 139
The source code — 139
Summary — 143

Chapter 5: Saga Snippets — 145
Source code overview — 146
Sample e-mail saga notification — 146
Using XAML — 148
The saga project — 150
Testing the solution — 155

Sample SFTP saga	**158**
Using XAML	160
Changing the process of messaging	160
Setting up an SFTP test environment	161
Saga deployment	**165**
ActiveMQ	**168**
The source code	168
Summary	**172**
Chapter 6: Using NServiceBus in the Cloud	**173**
Introducing the cloud and NSB	**173**
Introducing PaaS, IaaS, and SaaS	**175**
Using Microsoft Azure	**177**
Introducing Azure Storage Services	178
Azure Service Bus and Storage Queues	180
Azure Storage Queues and NSB	181
Azure Service Bus in NSB	185
Summary	**193**
Index	**195**

Preface

NServiceBus (NSB) is the most popular Enterprise Service Bus (ESB) in C#. It complements many of the other C# frameworks by providing an end-to-end ESB framework solution to work with services and messaging. The website, `http://particular.net/`, has many tools to assist in building endpoints, services, and messaging in Visual Studio. Visit `http://particular.net/downloads` for more details. There are also production tools to check on heartbeats of running endpoints and provide deep insights into running endpoints, services, and messages. NSB provides rapid development to allow integration into many different endpoints and services, for instance, e-mail, Secure File Transfer Protocol (SFTP), and the Windows Communication Foundation (WCF) protocol for web services.

Let's discuss using NHibernate, an object-oriented mapper (ORM) that maps objects to SQL databases, such as MySQL and SQL Server. As a developer, you will need to provide the mapping interface, usually an hbm.xml interface. While creating endpoints and sending messages NSB takes care of the mapping interface. This includes the creation of tables, logging, message durability, message retries, encryption, and many more components that help ensure high quality of software with the use of NSB. NSB provides many components, unique to NSB, needed for automation. NSB provides the following advantages:

- Separation of duties: There is separation of duties from the frontend to the backend, allowing the frontend to fire a message to a service and continue in its processing, not worrying about the results until it needs an update. Also, the separation of workflow responsibility exists through separating out NSB services. One service could be used to send payments to a bank, and another service could be used to provide feedback of the current status of the payment to the MVC-EF database, so that a user may see their payment status.

- Message durability: Messages are saved to queues between services so that if services are stopped, it can start from the messages in the queues when it restarts, and the messages will persist until told otherwise.

- Workflow retries: Messages, or endpoints, can be told to retry a number of times until they completely fail and send an error. The error is automated to return to an error queue. For instance, a web service message can be sent to a bank, and it can be set to retry the web service every 5 minutes for 20 minutes before giving up completely. This is useful during any network or server issues.
- Monitoring: NSB ServicePulse can keep a finger on the pulse of its services. Other monitoring can easily be done on the NSB queues to report on the number of messages.
- Encryption: Messages between services and endpoints can be easily encrypted.
- High availability: Multiple services or subscribers could be processing the same or similar messages from various services that are living on different servers. When one server, or service, goes down, others could be made available to take over those that are already running. Sagas are at the heart of the NServiceBus (NSB) workflow. Sagas save the message state in the form of saga data. They can retrieve the data as it is related to a message to update the message or data. This allows the NSB workflow to control the flow of data and messaging from end to end and correlate saved data to messages in the saga.

Messages are the means to transfer interaction and data between services in a service-oriented architecture (SOA). Sagas correlate, save, route, and manage processes that are started by these messages between services. Sagas even provide timeouts to ensure that messages do not live forever in a system.

Sagas provide decoupling between frontend websites and backend processes. It allows workflow to transfer so that a website can continue to do its work without having to wait for processes to return, such as the dreaded message "Please do not refresh this website as we bill your credit card".

NSB doesn't stop at development on Windows servers and desktops, but plays a big part in cloud development, for example with Microsoft Azure and the Microsoft cloud. NSB uses Service Bus as well as storage queues, SQL Server, and other storage containers. As the cloud is used more and more, NSB plays a key part in Software as a Service, as it is the premier framework for ESB in the C# clouds.

Even if your cloud solution doesn't end up being a C# compatible cloud, as many might hide the software running behind the cloud as being preparatory, NSB is a component for those that will do hybrid solutions, such as keeping data resident on-site. The connection to the cloud then will likely be via web services, and NSB sagas might likely provide the workflow to those web services.

From this book, you will discover the many features and characteristics of NSB, as follows:

- Sagas handle messages. A saga is started, or updated, by a message and passes it through its message handler. As messages are passed into the saga, the saga updates its sagas data from these messages through a message handler. The message logic doesn't normally end at the saga, but the saga creates a new one and passes it on to the next service. The saga may also respond back to the originating client. The saga routes messages while saving state and may be routing based on the previous state.

- Sagas contain long-lived transactions (LLTs) that contain database information for the messages for relatively long periods of time. An LLT is used when conditions such as short-lived transactions are not adequate. A short-lived transaction occurs when a call to a database, or MSMQ, performs a straightforward rollback or commit. For queues, NServiceBus performs second-level retries (SLRs) to try to commit a message a number of times before performing a rollback. In LLTs, there can be multiple conditions and multiple actions that need to take place for a message to be fully completed, or else operations execute the message from the beginning.

   The message is changed from one type of message into another, as one is handled by the saga, and the saga may create a new one with the same ID to pass to another service. Even though the message is different, it is a continuation of the flow of the original message that is considered a single transaction. The transaction is the accumulation of messages as they flow from one end to another end with the same message ID so that it represents the same transaction. The messages may be a different message type as it passes through different services. For instance, it may be an order message for an order service. The transaction can take seconds, days, hours, or longer, as the services take responsibility for acting upon it.

- Sagas contain timeouts for timing out messages and states. Because messages can be long-lived, services are responsible for retrieving and moving them. Sagas can have timers set on messages and data so that it doesn't live forever or the timer could be part of the business logic; for example, a customer has 30 seconds to enter a pin for their debit card. Sagas contain state information. Sagas save saga data to the database based on the message's data. The saga data is initially started with a message, and it is also updated with messages that are passed in with the same identification information. When a message passes between different services, saving the state information before the next service is wise, as there might be a business requirement to revert to its original state.

Preface

NServiceBus is the C# platform of choice for those that require workflows. In sagas, high availability, high performance, monitoring, encryption, rapid deployment, and many more features can only be found in this framework when building C# solutions.

What this book covers

Chapter 1, Introduction to Sagas, discusses NServiceBus and a basic design pattern that it uses, known as sagas, which is used to save states of messages. We will discuss the benefits of sagas and what it brings to the table with regard to software design.

Chapter 2, NServiceBus Saga Architecture, expands on the uses of sagas for persistence, timeouts, message durability, and message handling. We will discuss various message exchange patterns through examples to include gateway and cluster managing. These are important concepts as they drive the high availability and high performance that NSB brings to the table.

Chapter 3, The Particular Service Platform, is an overview of the Particular website-associated tools for NServiceBus. We will discuss building sagas through the ServiceMatrix tool, which is a Visual Studio extension tool for visually designing NSB endpoints, messages, and services. The other tools that we will discuss at length as they apply to sagas are the ServicePulse to monitor the endpoint availability during production, and ServiceInsight to take a deep dive into the functionality and properties of endpoints, services, and messages as they execute.

Chapter 4, Saga Development, focuses on various useful constructions of sagas and message handlers. The purpose of sagas will be discussed as the discussion goes into the need for extending and coordinating transactional integrity using sagas. The chapter then morphs into a discussion of NServiceBus, using integrated pre-built WCF bridges. While some might consider it unusual to discuss WCF in a saga chapter, sagas become an intermediate for coordinating WCF and NServiceBus work. We can decouple the workflow from the frontend for interaction with backend processes through message handling. Sagas provide the means to persist the state information of the messages. This discussion will also handle other queuing sources as well to include RabbitMQ and ActiveMQ.

Chapter 5, Saga Snippets, discusses two primary saga examples, one using an e-mail, and one using the Secure File Transfer Protocol (SFTP). These samples sill demonstrate the saga workflow and the use of timeouts more in depth. The saga code will be a mediator between a frontend Windows Presentation Framework (WPF) and a backend client executing either an e-mail or SFTP. Using a saga as a mediator between frontend and backend code that will interface into an external server, there will be many added benefits and features. The external server interface, such as an e-mail server or SFTP server, is usually beyond our control and is in the control of external operations or organizations, such as a bank. So, the interface into these servers is all that we have to work with, and as business, software and operational needs increase, we need a framework that is robust enough to meet these demands. Thus, we have NSB and sagas.

Chapter 6, Using NServiceBus in the Cloud, gives an introduction to the cloud with a deeper dive into the Microsoft Azure cloud. The Azure Storage containers and Service Bus will be discussed at length. An Azure Storage example will be discussed, which will work on-premise using the Azure SDK and Azure Storage Emulator. Another example will be given with NSB sagas as it works through Service Bus in the Azure cloud.

What you need for this book

We will discuss and build many examples in this book using Visual Studio 2012. Visual Studio Express could be used to walk through the samples as well. The user is expected to have a very basic understanding of C# and Visual Studio as we examine the use of NServiceBus.

All of the examples were originally built in the Windows Server 2008 operating system, and tested in both Windows 8.1 and Windows Server 2012.

For many of the database pieces, SQL Server Express, version 2008, was used. NServiceBus and many Particular tools will have to be installed from http://particular.net/downloads. When installing NSB, DTC, MSMQ, and RavenDB, performance counters will have to be installed. This book will walk you through those steps.

Preface

Who this book is for

This book is for the beginner or intermediate C# developer who wants to learn how to develop with NServiceBus and explore the NSB sagas, which is the workflow heart of NSB. Many beginning pieces will be discussed with a deep dive into sagas using many different Microsoft frameworks, which will also provide some basic knowledge of using WPF, WCF, ASP.NET MVC, and Entity Frameworks.

Conventions

In this book, you will find a number of text styles that distinguish between different kinds of information. Here are some examples of these styles and an explanation of their meaning.

Code words in text, database table names, folder names, filenames, file extensions, pathnames, dummy URLs, user input, and Twitter handles are shown as follows: "The `RequestTimeout()` function could take seconds, minutes, hours, days, or could be executed at a specific time of day."

A block of code is set as follows:

```
using System;
using System.IO;
using ServiceControl.Plugin.CustomChecks;
using ServiceControl.Plugin.CustomChecks.Messages;
using ServiceControl.Plugin.CustomChecks.Internal;
namespace PaymentEngine.ECommerce
{
public class MyCustomCheck : CustomCheck
{
public MyCustomCheck()
: base("ECommerce SubmitPayment check", "ECommerce")
{
ReportPass();
}
}}
```

Any command-line input or output is written as follows:

PM> Get-NserviceBusLocalMachineSettings

New terms and important words are shown in bold. Words that you see on the screen, for example, in menus or dialog boxes, appear in the text like this: "When the `TimerSubmit` process receives the reply message, it will pop a **Timer in Seconds** window as follows."

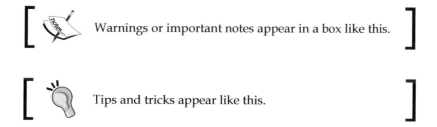

Reader feedback

Feedback from our readers is always welcome. Let us know what you think about this book—what you liked or disliked. Reader feedback is important for us as it helps us develop titles that you will really get the most out of.

To send us general feedback, simply e-mail `feedback@packtpub.com`, and mention the book's title in the subject of your message.

If there is a topic that you have expertise in and you are interested in either writing or contributing to a book, see our author guide at `www.packtpub.com/authors`.

Customer support

Now that you are the proud owner of a Packt book, we have a number of things to help you to get the most from your purchase.

Downloading the example code

You can download the example code files from your account at `http://www.packtpub.com` for all the Packt Publishing books you have purchased. If you purchased this book elsewhere, you can visit `http://www.packtpub.com/support` and register to have the files e-mailed directly to you.

Errata

Although we have taken every care to ensure the accuracy of our content, mistakes do happen. If you find a mistake in one of our books—maybe a mistake in the text or the code—we would be grateful if you could report this to us. By doing so, you can save other readers from frustration and help us improve subsequent versions of this book. If you find any errata, please report them by visiting http://www.packtpub.com/submit-errata, selecting your book, clicking on the **Errata Submission Form** link, and entering the details of your errata. Once your errata are verified, your submission will be accepted and the errata will be uploaded to our website or added to any list of existing errata under the Errata section of that title.

To view the previously submitted errata, go to https://www.packtpub.com/books/content/support and enter the name of the book in the search field. The required information will appear under the **Errata** section.

Piracy

Piracy of copyrighted material on the Internet is an ongoing problem across all media. At Packt, we take the protection of our copyright and licenses very seriously. If you come across any illegal copies of our works in any form on the Internet, please provide us with the location address or website name immediately so that we can pursue a remedy.

Please contact us at copyright@packtpub.com with a link to the suspected pirated material.

We appreciate your help in protecting our authors and our ability to bring you valuable content.

Questions

If you have a problem with any aspect of this book, you can contact us at questions@packtpub.com, and we will do our best to address the problem.

Introduction to Sagas

In this chapter, we will discuss an introduction to sagas. Sagas are just one of the many design patterns supported by **NServiceBus** (**NSB**). NSB is an **Enterprise Service Bus** (**ESB**) that brings many design features and benefits to the end-to-end enterprise. We will briefly look at these features and enter into a brief discussion of some of the design patterns that NSB supports. From this point, we will introduce the saga design pattern and provide a brief introduction to its purpose. The rest of the book will lead into a drill-down of the saga pattern and its use, as well as discuss many more of the features of NSB.

In this chapter, we will study the following:

- A brief introduction to ESBs
- Additional patterns
    - The publish-subscribe pattern
    - The request-reply pattern
    - The DataBus pattern
    - Timeout patterns
    - Message mutation patterns
    - Message encryption patterns
    - The ScaleOut pattern
    - The saga design pattern
- Sagas – what are they good for?

A brief introduction to ESBs

NServiceBus is an Enterprise Service Bus. An ESB, a software architecture model, is usually a framework that is used for designing and implementing the end-to-end pieces of **service-oriented architecture (SOA)**. SOA is a software architecture model based on distinct pieces of software, called services, providing application functionality as services to other applications. ESBs are common methods for designing SOAs. NSB is the most popular C# ESB. An ESB is a bus that brings a common communication mechanism together between services. Services are an abstract concept of a managed, self-contained process with messaging used to talk to the loosely-coupled, asynchronous, message-exchanging, monitored, managed, scalable, reliable, durable, and standard services.

SOA principles are widely implemented in the industry in the form of web services as web service messaging, usually across HTTPS sending XML-like messaging defined in **Web Service Definition Language (WSDL)**. ESBs, implementing SOAs, decouple the frontend services from backend services. Some of the benefits of decoupling services include the ability to work on the services dependently and separately, which includes adding updates to one without affecting the others. The interfaces between the services are messages, usually in XML, that define the data to be exchanged between messages. An example of how decoupling is useful is: updating interfaces of banks and accounting systems may need to be performed, and during these updates, the working of frontend is totally autonomous to the services that are affected. Decoupling expedites testing, maintenance, reduction of coding side-effects, and assists in breaking down business logic into discrete pieces as services.

ESBs extend the use of SOAs, as they provide a common communication medium, known as a bus, to transmit messages in a managed form that is centrally monitored. Unlike a **Service Broker**, the ESB ensures message integrity so that the message arrives correctly and the transaction is centrally managed. ESBs can use multiple means to communicate messages, even in the form of web services, but they use a common framework to monitor the endpoints, messaging, and services. Otherwise, the pieces are more desperate and cannot be used collectively. Depending on the ESB used, it may collectively manage these pieces. ESBs are also event driven and rely on message queuing, so that when the bus receives a message from a service, it routes it to the next appropriate service to process the business logic. Here's a common bus in an ESB, managing messages to different endpoints:

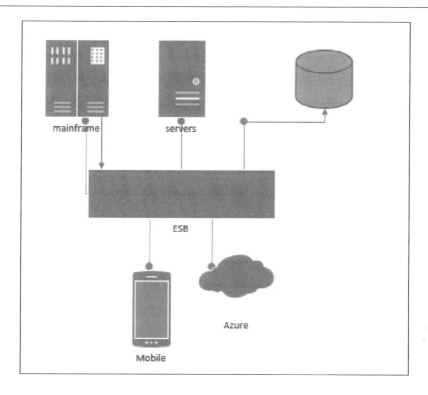

The preceding diagram show very high-level interfaces, so we can drill down into the protocols of the services themselves.

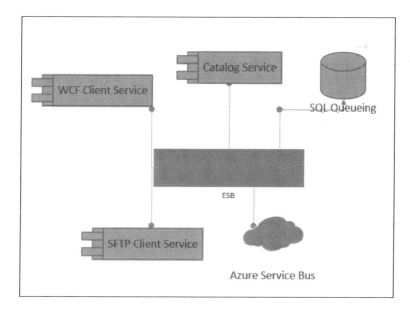

Introduction to Sagas

By using NSB, we can rapidly develop end-to-end applications in Visual Studio. There are also tools to deploy services, monitor all endpoints, and check the integrity of messages from end to end to ensure proper SOA. ESBs have many pieces of cross-cutting functionality that improve the quality of software.

The purpose of the cross-cutting quality of software is to ensure maintainability, security, high availability, reporting, alerting, scalability, performance, and other operational integrity features. An ESB such as NSB helps by sharing the common framework for a service bus. For instance, I may want to encrypt only the SSN of a message from end to end. By using the same base code and framework, we can encrypt and decrypt at various endpoints' SSN pieces with the same codebase.

By using a framework such as NSB, we achieve unity in cross-cutting functionality. The use of NSB's service bus technology can bring many disparate communication protocols to act with a common methodology and workflow to use all the benefits of NSB, for the use of endpoint protocols such as the **Secure File Transfer Protocol** (**SFTP**), Microsoft Web stack for more Restful Web APIs, and Microsoft's WCF for web services. The queuing mechanisms, which are also endpoints, consist of Azure Services Bus or Azure Storage Queues, communicating through SQL queuing, **Microsoft Message Queuing** (**MSMQ**), ActiveMQ™, and RabbitMQi. There are many endpoints, from queuing services to third-party vendors, which can be used in NSB.

Another instance of cross-cutting functionality is using the same code. Here are some of the benefits of using an ESB, such as NSB:

- **Message durability and error handling**: Messages are guaranteed to be delivered. If an error occurs during message delivery, it can roll back and place the message in the error queue. There are first-level and second-level retries to get the message to work successfully.

- **Message queuing and fault tolerance**: Messages persist in a queue until they are handled. If there is something wrong with the service, the message will remain in a persisted queue until the service is working successfully. No message will ever be lost.

- **High availability and high performance**: Services can be cloned and clustered as needed to ensure bandwidth utilization of messages and services. These clones can be coded and configured based on meeting the **service-level agreements** (**SLAs**).

- **Extremely configurable**: NSB is configured using the `Configure.With()` function that can easily change the persistence from RavenDB to the NHibernate ORM framework, or queuing from MSMQ, RabbitMQ, ActiveMQ, or SQL Queuing. Most of the code, if not all, except some configurations, will work the same regardless of the persistence and queuing models, as NSB takes care of the endpoint provisioning and mapping.
- **Service hosting**: NSB takes care of the management and hosting of services using the `NServiceBus.Host.exe` file.
- **Central production monitoring**: Particular's ServicePulse can monitor the production endpoint, message reporting, and service reporting.
- **ServiceInsight**: Particular's ServiceInsight can provide a deep insight through visual canvases, endpoint views, message views, and a deeper understanding of saga's messages, endpoints, and services.
- **Rapid development**: Particular's ServiceMatrix is a development studio extension for Visual Studio to provide rapid development through graphical canvases that generate C# code.

Let's look at a couple of practical needs for ESB designs.

- **Payment engine to a bank**: A C# programmer could use the Microsoft **Windows Communication Foundation (WCF)** or Microsoft Web APIs to perform web services for an external bank to debit a credit card. There are many pieces that WCF will not perform to support cross-cutting concerns, such as retries and failures. NSB can host the service and provide durable messaging in MSMQ that performs message retries on the WCF interface. NSB will persist with messages if there is an interruption of web services to third-party services, such as a bank. This includes keeping track of message reporting and even the encryption of messages. So, while WCF can implement the web service piece, there are many pieces outside of the realm of WCF that only a product like NSB can provide cross-cutting functionality for.
- **An online ordering system**: When ordering a product, you may be greeted with a page that spins and states **Do not refresh, processing order**. This is clearly a design where the website is tightly coupled to the ordering site, including the processing of the credit card. There are many automated processes that may disturb a web browser, or there may be some disconnect on the server side that may start the process over again. An ESB will always ensure that if you have to place your order again or had an interruption, your messaging is transactional and durable, ensuring integrity when charging for your order.

> For a website that is loosely coupled with an NSB design, the website could just send a message to the service queue without interrupting the web flow. A site that doesn't allow refreshing when processing an order depends on backend processes to finish, where the web flow may be interrupted. An NSB design can still give the user feedback on the status of their order without any interruption to the website.

Event-driven jobs

For scheduling jobs on Windows, the Windows Task Scheduler can do the job (see http://en.wikipedia.org/wiki/Windows_Task_Scheduler). However, depending on the complexity of the job, the Windows Task Scheduler may not be enough to deal with it. For instance, if a holiday schedule needs to be checked before starting the job and if the job has dependencies on other events and services happening before it runs, you may run into limitations with the task scheduler. The Windows Task Scheduler is a great tool, but it is not event driven, as is an ESB. With ESBs, time is an event.

Even though a job may be easily done by Task Manager, you will not get the global management of services, the messaging of events, commands, and messages, and the feedback of the processing from a managed perspective, unless you use a framework such as NSB.

NSB extends the design to allow many facets of event-driven designing where building a simple website, scheduling a daily job, or sending payments can easily grow into something much larger in order to have, monitor, manage, extend, secure, and maintain great software quality. It is not uncommon that when these simple designs become complex, a rewrite is warranted. That is simply because software quality and extending the design were not taken into consideration ahead of time.

The ESB is an architectural software design pattern and framework for designing SOA architectures. However, there are many additional software design patterns that an ESB can bring to the table as well. These will be software design patterns that NSB supports, but not all ESBs support these patterns.

Additional SOA patterns

There are many additional software design and messaging patterns that NSB supports. Many such patterns can be found at `http://www.enterpriseintegrationpatterns.com`. The purpose of design patterns is to provide reusable frameworks that have been hardened through reuse in the industry, to rapidly deploy what's been designed before. NSB has several design patterns that are popular in any robust and solid ESB solution.

The publish-subscribe pattern

One of the biggest benefits of using ESB technology is the availability of the publish-subscribe message pattern (for more details, see `http://en.wikipedia.org/wiki/Publish-subscribe_pattern`). The publish-subscribe pattern has been around for a very long time, and its simplicity is the cornerstone of many, many systems.

The pattern makes use of message queues, and NSB stores its messages through the use of queues, be it MSMQ, SQL queues, RabbitMQ, Azure queues, or ActiveMQ. The basic premise is that a message is published to a queue; in NSB, it uses the `Bus.Publish(message)` function that places an event message in a queue that does not need to be sent to specific receivers. There can be multiple output channels called subscribers listening in on the publisher. The benefit is that the services do not depend on any relationship with each other except the message, and can act independently of each other. This is a service which is decoupled. This means that dependencies are separated from each other.

Of course, normal NSB features still apply. For instance, the message will stay in the queue until the subscribers pull the message off the queue to handle it. For NSB, this means that the machine can be rebooted and the message will still be available in the queue until the subscribers process the message, which ensures that a message is never lost.

The subscribers will process the message, and there could be one or many subscribers. For instance, the publisher may put 10 different types of messages in its queue, and there could be a subscriber for each type of message.

The NSB framework has plugins for Windows Performance Monitor. In Service Level Agreement (SLA) code of the endpoint, we can specify that new services will be created or alerts will be sent if the service does not meet performance criteria. So if one of the subscribers is not performing well, it will force the other subscribers to take over, as they are performing better across multiple machines. Because of this, NSB is a highly available, durable-messaging, high-performance ESB engine that can be developed in a multitude of ways. A design of the publish-subscribe pattern appears as follows:

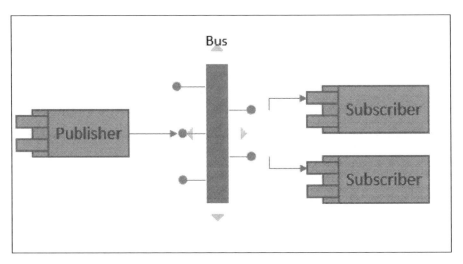

In NSB, there are special types of messages used specifically for publish-subscribe, called events. This forces the code to handle the message in a publish-subscribe scenario, and not in a request-reply scenario.

The publisher-subscriber mappings are done through the application configuration file. This will define the type of messages from the publisher queue that it will subscribe to, to handle the messages. When these programs execute, the subscription mappings will be further saved into a local database. By default, they are saved in RavenDB.

The request-reply pattern

The request-reply pattern is different from the publish-subscribe pattern as it sends a message to a directed endpoint.

> Please see http://en.wikipedia.org/wiki/Request-response.

The NSB framework will use the `Bus.Send(message)` function for request-reply. The message remains durable as it is still queued. The sender may or may not be available to handle the response, and the replier may not send a response, but the functionality in NSB is available to have these components easily added to handle the request-reply on their endpoints. The sender could easily be a website and the replier could send a response to update the website, for instance, when a credit card has been processed. The request-response pattern doesn't require subscription information to be stored, but it does require the endpoints to be defined along with the message types in the application configuration file. Sagas automatically store the sender's original information and reply directly to the requester, without storing extra information.

This pattern still has high availability and high performance, and still uses a multitude of topologies with message durability. Please see the scale out pattern in this chapter on how request-reply will cluster across machines.

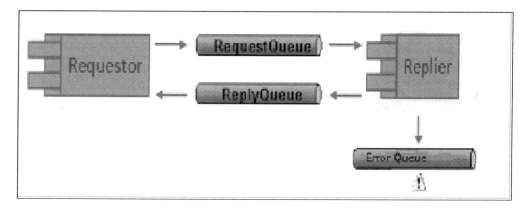

The gateway pattern

There are cases when services may be partly stored on one part of an organization's LAN, and other services are stored on another LAN, and the only transport to each other is an HTTP or HTTPS tunnel to pass messages to NSB.

The main purpose of the gateway is to allow you to do the same durable fire-and-forget messaging that you are accustomed to with NServiceBus across physically separated sites, where *sites* are locations in which you run IT infrastructure and not websites.

The gateway only comes into play where you can't use normal LAN-to-LAN VPN tunnels or use internal LAN servers to communicate MSMQ to MSMQ. The purpose of the gateway is to create messages that communicate through HTTP, but it would be preferable to use HTTPS to ensure that messages are secured. The following architecture diagram represents a gateway built using the NSB:

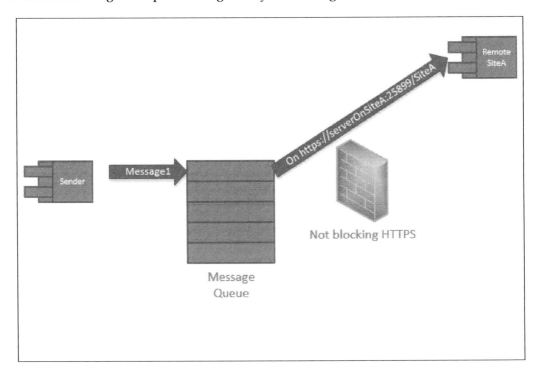

The source code

In this section, we will be using the following gateway solutions:

- `Headquarter.Messages`: This refers to the common messages for `Headquarters`, `SiteA`, and `SiteB`.
- `Headquarter`: This will receive messages from `http://localhost:25899/Headquarter/` and `http://localhost:25899/Headquarter2/`, and send messages to `http://localhost:25899/SiteA/` and `http://localhost:25899/SiteB/`.
- `SiteA`: This is a project that will receive updated price information from `Headquarters` across `http://localhost:25899/SiteA/` and respond that it was successful to `Headquarters` across `http://localhost:25899/Headquarter2/`.
- `SiteB`: This is a project that will receive updated price information from `Headquarters` across `http://localhost:25899/SiteB/`.
- `WebClient`: This will have an `Index.htm` page to send a JSON script to `http://localhost:25899/Headquarter/`.

The preceding code solution is built using Visual Studio 2012 in Windows Server 2012, with MSMQ, DTC, NServiceBus references, and SQL Server 2012 Express LocalDB installed.

In a gateway, there are incoming channels and defined site keys to send outgoing messages to their sites. We can see in the application configuration file of the headquarters that the receiving channels for the headquarters are `http://localhost:25899/Headquarter/` and `http://localhost:25899/Headquarter2/`.

There will be a set of site keys for sending sites that make up `SiteA` and `SiteB`.

```
<GatewayConfig>
  <Sites>
    <Site Key="SiteA" Address="http://localhost:25899/SiteA/"
      ChannelType="Http" />
    <Site Key="SiteB" Address="http://localhost:25899/SiteB/"
      ChannelType="Http" LegacyMode="false" />
  </Sites>
  <Channels>
    <Channel Address="http://localhost:25899/Headquarter/"
      ChannelType="Http" />
    <Channel Address="http://localhost:25899/Headquarter2/"
      ChannelType="Http" Default="true" />
  </Channels>
</GatewayConfig>
```

Introduction to Sagas

The `Bus.SendToSites(new[] { "SiteA", "SiteB"})` allows you to send messages to multiple sites as you can see in the preceding configurations. For instance, the parameter of `SiteA` will send the message to `http://localhost:25899/SiteA/`.

Going across alternate channels such as HTTP means that you lose the MSMQ safety guarantee of exactly one message. This means that communication errors resulting in retry can lead to receiving the same message more than once. To avoid burdening you with de-duplication, the NServiceBus gateway supports this out of the box. You just need to store the message IDs of all received messages so it can detect potential duplicates. De-duplication tables can be stored on the SQL Server using the NHibernate persistence configuration. This will be configured on the IBus using the `.UseNHibernateGatewayDeduplication()` method. Of course, settings always need to be applied in the `App.config` file to define the database connection. Here, we are connecting to the local `SQLExpress` instance.

```xml
<connectionStrings><add name="NServiceBus/Transport"
    connectionString="cacheSendConnection=true"/>
  <add name="NServiceBus/Persistence" connectionString="Data
    Source=.\SQLEXPRESS;Initial Catalog=nservicebus;Integrated
    Security=True"/>
</connectionStrings>
<!-- specify the other needed NHibernate settings like below in
  appSettings:-->
<appSettings>
  <!-- dialect is defaulted to MsSql2008Dialect, if needed
    change accordingly -->
  <add key="NServiceBus/Persistence/NHibernate/dialect"
    value="NHibernate.Dialect.MsSql2008Dialect"/>
  <!-- other optional settings examples -->
  <add key=
    "NServiceBus/Persistence/NHibernate/connection.provider"
    value="NHibernate.Connection.DriverConnectionProvider"/>
  <add key=
    "NServiceBus/Persistence/NHibernate/connection.driver_class"
    value="NHibernate.Driver.Sql2008ClientDriver"/>
</appSettings>
```

This gateway pattern allows us to pass messages through HTTP/HTTPS to allow queuing across outside systems from the local LAN. Normally web services run through HTTP/HTTPS, but with NSB, the queuing process can also run through HTTP/HTTPS to outside servers, where all ports are blocked and only HTTP/HTTPS can be used.

The DataBus pattern

The DataBus is used to send large chunks of data or files across as an attachment because MSMQ is limited to 4 MB. For this reason, a reference can be passed on a local file to transfer data using the `databus` method. The message will provide a reference to a larger data file to be accessed that exceeds the message queue size due to size constraints.

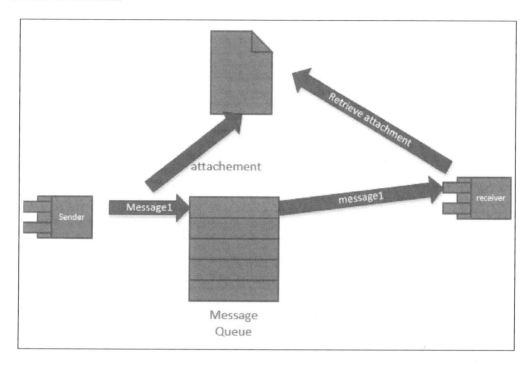

In this section, we will use the gateway solution.

The path of the data bus has to be set in the configuration of the endpoint. We will be using a relative path to where the gateway project is running. Both `SiteA` and `SiteB` will also have relative paths. There will be a relative path to the binary file with a data bus subdirectory containing the files that will have a lot of data. The following is the code:

```
public void Init()
{
    Configure.With()
        .DefaultBuilder()
        .FileShareDataBus(".\\databus");
}
```

Introduction to Sagas

When we execute the gateway project, we can mock the `SomeLargeString` variable to simulate data larger than 4 MB, as shown in the following code:

```
Bus.SendToSites(new[] { "SiteA", "SiteB" }, new PriceUpdated
{
    ProductId = 2,
    NewPrice = 100.0,
    ValidFrom = DateTime.Today,
    SomeLargeString = new DataBusProperty<string>("This is a random large string " + Guid.NewGuid())
});
```

If we execute the gateway project, it will create a message to the relative path of its binary file, save the message under data bus, and use it as a reference to send to `SiteA` and `SiteB`. Here, we see the message saved to the local relative path.

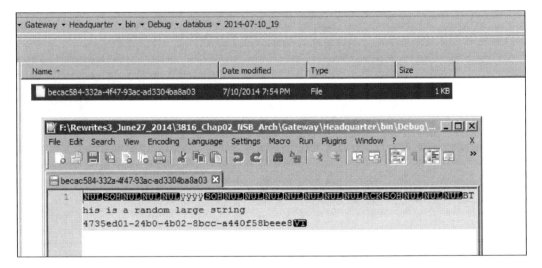

The data bus is very useful for processing files or chunks of data that are too large for MSMQ.

Timeout patterns

In ESB systems, the need for timers and timeouts cannot be underestimated. Many developers use the **Microsoft Task Scheduler**. The Microsoft Task Scheduler looks like this:

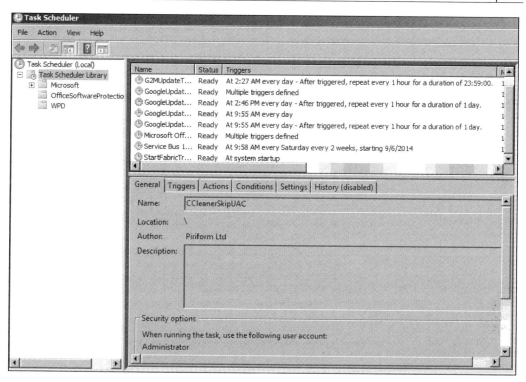

In NSB, we have the ability to use event-driven timer messages. This is the ability of a saga to start a function, run a command, or perform many other tasks, based on its timer. In the event-driven timer function, we can set a timeout for any time in the day, from seconds after the process starts, to minutes, hours, days, and even use a holiday table for the process. By having a managed service, the usage of timers can be more complex, for instance, adding holiday tables, not executing timers on weekends, and more business functions that task schedulers cannot handle. Also included in NSB, is the installation and management of the service itself. Time saved just in deploying services and having NSB set up DTC for the administrator may already pay for the NSB licenses.

The source code in this section will be in NSB Version 5.0 in the `TimerSaga - v5` directory. Here we have the following projects:

- `AppCommon`: This contains the ViewModel and context for the Windows forms.
- `TimerSubmit`: This is a project that submits a timer message to the saga between 1-100 seconds.
- `TimerMessages`: This contains common messages for the projects.

Introduction to Sagas

- `TimerSaga`: This saga will perform the timeout and respond back to the `TimerSubmit` project. This could easily start a cron job, execute a program, or direct other services just as easily as responding back to the submitting program.

In this solution, we have a Windows form, where we enter a variable from 1 to 100, based on which the `TimerSubmit` project will generate a message to the Timer saga to create a timeout for those seconds. After those seconds expire, the Timer saga will handle the timeout message and respond back to the `TimerSubmit` window saying that the timer has expired. This is a simple exercise in an event-driven timer that could have many uses.

The program will start by submitting the number of seconds between 1-100 to `TimerSubmit` as shown in the following screenshot:

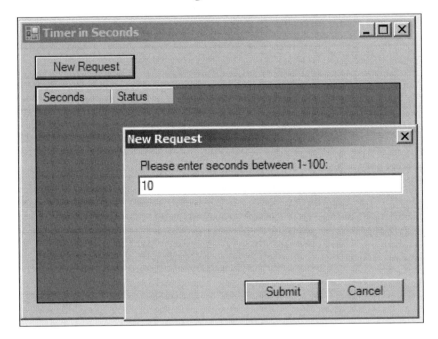

When we add a value between 1 and 100 seconds, the message will be sent to the queue of `TimerSaga` with the number of seconds and a `RequestId` to keep track of the message ID, as shown in the following screenshot:

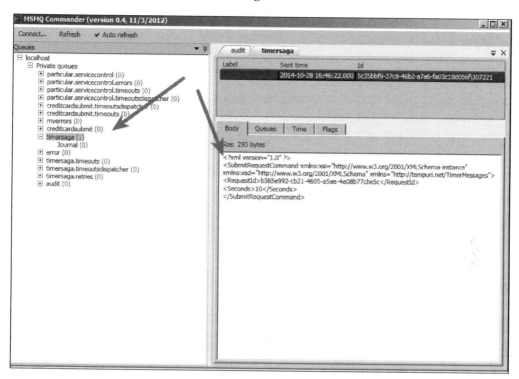

The message's `RequestId` is used to map the message to the saga data. The saga is started by `SubmitRequestCommand` with the number of seconds in it. It will save the saga data in the `TimerRequestData` object, which will allow the original client data to respond to it. The number of seconds will be set in the timeout, and when the timeout expires, it will execute a `TimeoutMessage` instance that will be handled by the saga. The saga's object class definition that defines this mapping appears as follows:

```
public class TimerRequestSaga : Saga<TimerRequestData>,
        IAmStartedByMessages<SubmitRequestCommand>,
        IHandleTimeouts<TimerRequestSaga.TimeoutMessage>
{
```

Introduction to Sagas

The starting message `SubmitRequestCommand` will be handled by the saga's message handler in the following code:

```
public void Handle(SubmitRequestCommand message)
{
    logger.Info("--------TimerSaga SubmotRequestCommand-------" + message);

    var command = new TimeoutMessage
    {
        RequestId = message.RequestId
    };

    // We put the number of seconds 1-100 in the timer
    RequestTimeout<TimeoutMessage>(TimeSpan.FromSeconds(message.Seconds),command);

    Data.RequestId = message.RequestId;
    Data.Seconds = message.Seconds;

}
```

This code will set the timer that will send a `TimeoutMessage` instance with the message's `RequestId`. All the messages that are part of the same starting message will have the same `RequestId`. This will keep track of which message started the next message, as well as map the saga data to save and retrieve it from the saga. The `RequestId` is handled like a primary key to the saga data that NSB will use to map the data to the messages. NSB handles the mapping, but we must define the unique identifiers and keep processing them in the messages. If this seems complicated, keep reading, as it is broken down in subsequent chapters.

When the timeout that we passed to the `RequestTimeout<>()` code is reached, an instance of a timeout message is passed back to the saga. If it was set for 10 seconds, after 10 seconds have passed, the following message handler in the saga is called to process the timeout message that we set in the timer. The `Timeout` message handler appears as follows:

```
public void Timeout(TimeoutMessage state)
{
    logger.Info("--------TimerSaga Timeout-------");
    logger.Info("Data.RequestId:" + Data.RequestId);
    var command = new SubmitRequestReplyMessage
    {
        RequestId = state.RequestId
    };

    ReplyToOriginator(command);
    MarkAsComplete();
}
```

This code will process the `TimeoutMessage` instance that we called `state`. We had mapped the `RequestId` in the version 5.0 Saga mapper code, `ConfigureHowToFindSaga(SagaPropertyMapper<TimerRequestData> mapper)` to map the message's `RequestId` to the data, so the `state.RequestId` will be the same as the `Data.RequestId` data instance of saga. The `RequestId` is checked to ensure that when we execute `MarkAsComplete()`, the correct saga store is deleted.

This can be broken down into simple steps:

1. A message is sent to the saga with a timeout value.
2. The saga values are stored and the timeout message is created, all using the same `RequestId`.
3. The timeout expires.
4. A timeout message is handled by the saga.

The saga data is deleted, and a message is sent to the originator saying that the timeout has been completed. In this message handler, we create a `SubmitRequestReplyMessage` instance with the `RequestId`, which will be the same `RequestId` as the message that started this process. We reply to the originator, which will send this message to the original `TimerSubmit` program.

Introduction to Sagas

When the `TimerSubmit` process receives the reply message, it will pop a **Timer in Seconds** window as follows:

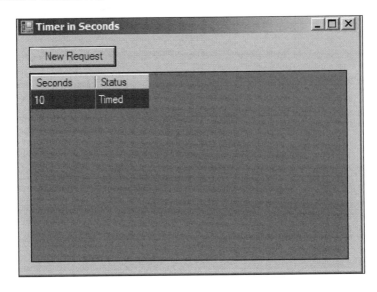

This code has all the pieces of a simple saga. The code could be created with a combination of other frameworks as well, such as Quartz for the timer and TopShelf to create services, but NSB is a complete framework that provides end-to-end pieces of architecture and many patterns that have all the pieces.

The end message that showed the **Timed** status in the Windows form could easily be the execution of a job, another program, an e-mail of system status, a report of the system status, and many more timed tasks. The number of seconds used earlier was just a demonstration. The `RequestTimeout()` function could take seconds, minutes, hours, days, or could be executed at a specific time of day. We could execute the function with a weekend and holiday calendar and extend the functionality further. We could create a centralized saga timer to literally schedule all the tasks like Microsoft Task Scheduler, but create a managed service that can be monitored, managed, perform reporting, and improve functionality in a much further detailed solution.

Message mutation patterns

Message mutators allow you to change messages by plugging custom logic into a couple of simple interfaces. For instance, you can encrypt a part or all of a message. The encryption message mutator is part of the NServiceBus library, and can be used at any time. You can intercept the incoming message and then mutate it before sending it as an outgoing message. This is the process of changing messages as they leave a client and enter a server.

The source code

In this section, we will be using the `MessageMutators` solution with the following projects:

- `Client`: The client will send messages to the server.
- `Server`: This will receive the mutated message.
- `Messages`: This is the message format being passed between client and server.
- `MessageMutators`: This project will contain the mutation code to compress and decompress the messages in `TransportMessageCompressionMutator.cs` and validate the message annotation in `ValidateMessageMutator.cs`.

The client and server needs to be running. The client will prompt to send a good or bad message. The good message is compressed so that it will pass the 4 MB MSMQ buffer size, as shown in the following screenshot:

The queue's data will be validated and compressed from the client before processing it in the MSMQ. This is shown in the following diagram:

Client → Send message → Validate (Outgoing) → TransportCompression (Outgoing) ---> To MSMQ

Then, the server will receive the message from MSMQ, but before processing it, this will decompress and validate the message. It will restore the message that the client mutated. This is shown in the following diagram:

MSMQ → TransportCompression (Incoming) → Validate (Incoming)

This is just a simple compression and data annotation validation to ensure that MSMQ will process the message. One of the reasons for mutating the message may be the encryption of a credit card within a payment message.

Introduction to Sagas

Message encryption patterns

NSB supports the AES (Rijndael) encryption algorithm. AES stands for **Advanced Encryption Standard**. This is a symmetric key algorithm, so both the program encrypting data and decrypting data must share a secret key for their functioning. Visit http://en.wikipedia.org/wiki/Advanced_Encryption_Standard for more information on AES.

Encrypting data will depend on the needs of the organization, but common items include passwords, financial information, or personal customer identification information. AES is the strongest symmetric encryption algorithm, and most languages, such as Java and C#, have API support for its use.

We know that part of the configuration on both sides will be a secret key.

The source code

In this section, we will be using the Encryption solution with the following projects:

- Client: This will send an encrypted credit card message to the server
- Server: This will receive the credit card message and decrypt it
- Messages: This is the message format being passed between client and server

Both, the client and server must be running. The client will have a prompt to send messages to the server, as shown in the following screenshot:

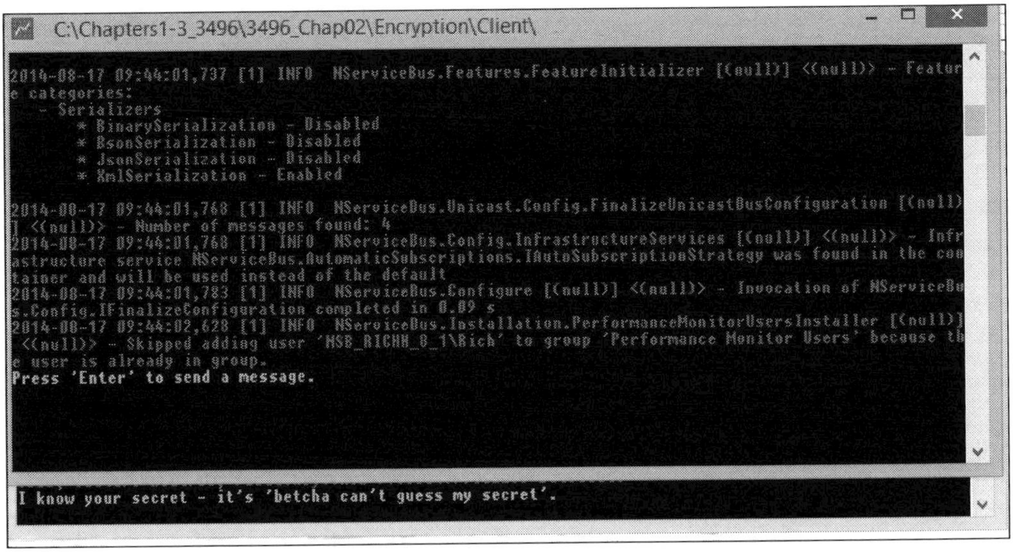

After pressing *Enter*, we see that the message is encrypted on the server queue:

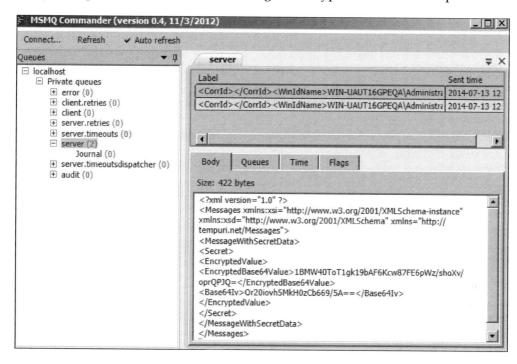

When running the server, NSB will decrypt the message before passing it to the server's message handler.

All that is really needed is to enable both ends for AES in the IBus by the configuration .RijndaelEncryptionService();. We set the part of the message that we want to encrypt by the public WireEncryptionString Secret { get;set; } where WireEncryptionString defines that the string will be encrypted. Also, the secret key has to be in the App.config file of both the client and server. The following screenshot shows the code:

```xml
<?xml version="1.0"?>
<configuration>
  <configSections>
    <section name="UnicastBusConfig" type="NServiceBus.Config.UnicastBusConfig, NServiceBus.Core"/>
    <section name="RijndaelEncryptionServiceConfig" type="NServiceBus.Config.RijndaelEncryptionServiceConfig, NServiceBus.Core"/>
  </configSections>
  <MsmqTransportConfig ErrorQueue="error" NumberOfWorkerThreads="1" MaxRetries="5"/>
  <UnicastBusConfig>
    <MessageEndpointMappings>
      <add Messages="Messages" Endpoint="Server"/>
    </MessageEndpointMappings>
  </UnicastBusConfig>
  <RijndaelEncryptionServiceConfig Key="gDbqRpqdRbTs3mhdZh8qCaDaxJXl+e7"/>
  <startup>
    <supportedRuntime version="v4.0" sku=".NETFramework,Version=v4.0"/>
  </startup>
</configuration>
```

Introduction to Sagas

The ScaleOut pattern

As mentioned earlier, one of the many benefits of using NSB is that you can distribute the load or the NSB services or processes. This is commonly known as scaling out the services. The idea is that you can deploy the same service across a farm or multiple servers. This is used to create an environment of high availability.

This model is a form of round-robin clustering, where a handler can distribute its workload to additional workers doing exactly the same work. A distributor is used with MSMQ. If an endpoint has a critical time set for performance and requires more processing help, this clustering is used to spawn off work to the same services living on other machines to share the load. If the machine processing the message crashes, the message rolls back to the queue and other machines can then process it.

Worker services send messages through a control queue saying that they are ready for work. The distributor stores these messages, and when it receives messages, it fetches them out of the available queues. All pending work stays in the distributor's queue so that messages can be timed for performance.

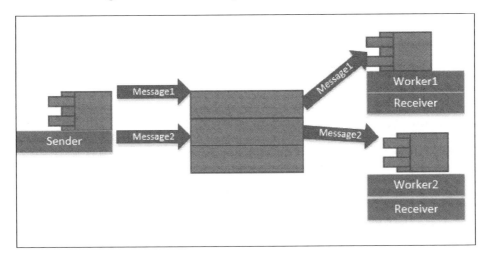

The saga design pattern

The saga design pattern starts with a message to the saga service. The saga service requires several components, as follows:

- `IAmStartedByMessages<IMessage>`: The saga's actions are started by an incoming message. This message initializes the saga data object and creates the first elements, which needs to include a primary key from this message for subsequent lookups. The saga data elements are created in the message handler. This is shown in the following screenshot:

Chapter 1

```
public class PaymentRequestSaga : NServiceBus.Saga.Saga<PaymentRequestData>,
            IAmStartedByMessages<SendCommand>,
            IHandleMessages<ResponseCommand>,
            IHandleTimeouts<SendCommand>
{
    public void Handle(SendCommand message)
    {
        ResponseCommand resp = new ResponseCommand();
        resp.RequestId = message.RequestId;
        resp.state = PaymentMessages.MessageParts.StateCodes.SentMyWCFClient;
        ReplyToOriginator(resp);
        Bus.Send(message);
        RequestTimeout(TimeSpan.FromHours(3), message); // 3 hours
        Data.RequestId = message.RequestId;
    }
```

- `Saga<IContainSagaData>`: The saga data is the data that is persisted. This data contains default values to be set by NSB for an ID, originator, and `OriginalMessageId` instance. Also, a unique ID should be used for the mapping of the message to data, and vice versa. This is shown in the following screenshot:

```
public class PaymentRequestData : IContainSagaData
{
    /***
     * Gets/sets the Id of the process. Do NOT generate this value in your code.
       The value of the Id will be generated automatically to provide the
       best performance for saving in a database.
     * ***/
    public virtual Guid Id { get; set; }  // Required
    /***
     * Contains the return address of the endpoint that caused the process to run.
     * ***/
    public virtual string Originator { get; set; }  //Required
    /***
     * Contains the Id of the message which caused the saga to start.
       This is needed so that when we reply to the Originator, any
       registered callbacks will be fired correctly.
     * ***/
    public virtual string OriginalMessageId { get; set; }  //Required

    [Unique]
    public virtual Guid RequestId { get; set; }  // Unique ID to lookup Request message
}
```

Introduction to Sagas

- `IHandleMessages<IMessage>`: The saga doesn't do much unless it is handling messages. It can retrieve the saved data of a message that is mapped by the configuration. NSB handles the mapping of the data to the message by the unique ID, in this case, `RequestId`, that is defined in `ConfigureHowToFindSaga()`.

- `ConfigureHowToFindSaga()`: This function of the saga pattern is used to map the data to the messages. Usually, the mapping can be performance from a GUID or ID, but it must be a unique data type that can be stored as a primary key in a database. The messages that are mapped contain the same key, and when it passes in the saga, the data is found that matches the key. This function is called when the saga object is instanced.

```
public override void ConfigureHowToFindSaga()
{
    ConfigureMapping<ResponseCommand>(x => x.RequestId).ToSaga(x => x.RequestId);
}
```

Sagas – what are they good for?

Sagas play a critical role in workflow management. Without the ability to map data to persistent data, workflow among messages cannot be achieved. Saga's purpose is to persist saga data objects, which are files with data from selected content, directed from messages in the message handler. The following are some of the things that sagas are good for:

- **Saving message state/session information**: When a message is started, the message handler will create the saga data instance that will be saved at the end of the message handler. It is up to the message handler data to set data into an instance of the object. NSB handles the mapping. The following screenshot displays the code for handling messages:

```
public void Handle(SendCommand message)
{
    ResponseCommand resp = new ResponseCommand();
    resp.RequestId = message.RequestId;
    resp.state = PaymentMessages.MessageParts.StateCodes.SentMyWCFClient;
    ReplyToOriginator(resp);
    Bus.Send(message);
    RequestTimeout(TimeSpan.FromHours(3), message); // 3 hours
    Data.RequestId = message.RequestId;
}
```

- **Changing message state/session information**: During a message handler, the `ConfigureHowToFindSaga()` function has mapped a way for NSB to retrieve the correct saga data object, based on the message's unique ID. In the initial message that started the saga, the initial data and ID is set to the saga data row. Subsequent messages will either update this row of data, retrieve it, or delete it. This data is the workflow used to store the state between messages.

```
public void Handle(ResponseCommand message)
{
    if (Data.RequestId != message.RequestId)
    {
        message.state = PaymentMessages.MessageParts.StateCodes.Fail;
    }
    ReplyToOriginator(message);
    MarkAsComplete();
}
```

- **Responding to originator/original client of changes**: The saga can respond to the original client with the `ReplyToOriginator()` function. This can complete the path of the original message with a success or error message to complete the end-to-end trip of the message. The `MarkAsComplete()` function states that the data object is no longer needed and can be deleted.
- **Providing timeouts**: Sagas have the ability to provide timeouts. There could be many reasons for timeouts, but the most obvious one is not having a message running that has associated saga data to run forever. The `RequestTimeout()` function can be a daily timeout, a one-time timeout, a timeout based on seconds, minutes, or hours, and other variations involving time. Other uses of timeouts could be an execution of jobs to run, more messages to send based on time, performing a timeout to update the originator of a status, and many more uses.

Introduction to Sagas

In this scenario, we give the message three hours to perform its operations before timing it out. The following screenshot displays the code for providing timeouts:

```
public class PaymentRequestSaga : NServiceBus.Saga.Saga<PaymentRequestData>,
        IAmStartedByMessages<SendCommand>,
        IHandleMessages<ResponseCommand>,
        IHandleTimeouts<SendCommand>
{
    public override void ConfigureHowToFindSaga()...

    public void Handle(SendCommand message)
    {
        // ....
        RequestTimeout(TimeSpan.FromHours(3), message); // 3 hours
        // ....
    }

    public void Handle(ResponseCommand message)...

    public void Timeout(SendCommand state)
    {
        ResponseCommand resp = new ResponseCommand();
        resp.RequestId = state.RequestId;
        resp.state = PaymentMessages.MessageParts.StateCodes.Timeout;
        ReplyToOriginator(resp);
        MarkAsComplete();
    }
}
```

Because the saga pattern has the ability to persist the message's state data, the messages become connectionless pieces of separate messages and provide a means to connect multiple messages to each other to perform end-to-end flow. Thus, the saga becomes the focal point of the interaction. The messages now have a behavioral pattern as they have more meaning than a single instance. They now have a long time to live, as when the individual message is no longer available, the saga data representing data for the message will still live on, until it is marked to be destroyed.

Summary

In this chapter, we introduced the need for the use of the saga design pattern as it applies to NSB. We discussed some of the basic features and uses of NSB, as an ESB framework for SOA development. It adds quality to software as a framework standard, with many extensions for logging, security, reporting, and message handling between services. The flow of NSB is the automation of the backend and the decoupling from the frontend, so that automation can occur with services that have message flow from end to end.

NSB supports many design patterns that make up the ESB. The saga pattern provides the means to persist message data, which allows for end-to-end workflow. It can easily respond to the originator of the message, thereby updating the status of an end-to-end flow. The saga data object saves the state of the message to allow services to change the state as the message changes when it flows from end to end in the NSB. NSB solves many design issues, including the need for high availability and behavior changes based on messages for the flow of the end-to-end design. We also discussed many more patterns.

In the next chapter, we will discuss saga architecture.

2
NServiceBus Saga Architecture

We have briefly touched upon sagas in the previous chapter, and in this chapter, we will go into more detail. We will discuss timeouts, message handling, and persistence for sagas. In this chapter, we will cover:

- Transition from NServiceBus (NSB) version 4 to version 5
- Message flow
- Deployment
- Insight

Upgrading from NSB version 4 to 5

In the later part of September 2014, NServiceBus introduced version 5.0, an upgrade from version 4.x. NSB version 5.0 requires different coding techniques. Currently, NSB is in version 5.02, which offers many features. These are as follows:

- Non-DTC operations: These support durable messaging in cloud queues and other queuing systems, such as RabbitMQ, which do not support local transaction management. Version 5 has been modified to provide additional support for integration into Windows Azure queues.
- ISendOnlyBus: This simplifies the ability to create a configuration just for the purpose of sending messages only to the bus. The function used will look like this: `var bus = Bus.CreateSendOnly(configuration)`.
- `Bus.Create(new BusConfiguration())`: The bus configuration in version 5.0 is different from version 4 as it supports the bus configuration object that is passed around through functions, as there can be different bus configurations used for different reasons.

- Move to .Net 4.5: Version 5.0 requires .NET 4.5 as a minimum requirement, which will ensure that NSB works best with packages that are supported in Visual Studio 2012 and above. This ensures that integration into Visual Studio 2012 packages and tools is optimized.

Thus, the minimum requirement for version 5.0 is Visual Studio 2012. The new minimum requirement for the server operating system (OS) is 2008 and the desktop OS is Vista SP2.

There are many more benefits of using NSB 5.0 over 4.0. However, the code used in version 4.0 will require changes to work properly with version 5.0. This is due to the configuration of the bus being instantiated and passed through some of the functions, as the bus is a more configurable object than a static element in the program. For a list of the features available during the upgrade, please see `http://docs.particular.net/nservicebus/upgradeguides/4to5`. For example, we can see the following code that was done in NSB version 4.0:

```
logger.Info("--------AppForAccountingDept IBus Config Start--------");


Configure.With()
    .UnityBuilder(container)
    .UseTransport<Msmq>()
    .DisableTimeoutManager()
    .UnicastBus()
    .CreateBus()
    .Start(StartupAction);

logger.Info("--------AppForAccountingDept IBus Config End--------");
```

The following is an example of a similar code done in NSB version 5.0:

```
logger.Info("--------AppForAccountingDept IBus Config Start--------");
var busConfiguration = new BusConfiguration();

// Unity with a container instance
busConfiguration.UseContainer<UnityBuilder>(c => c.UseExistingContainer(container));

busConfiguration.UsePersistence<InMemoryPersistence>();
busConfiguration.EnableInstallers();
Bus.Create(busConfiguration).Start();//this will run the installers

logger.Info("--------AppForAccountingDept Running--------");
```

These examples show that in version 5.0, the bus configuration is created as an object and configuration values are added to the instance of the object. The bus configuration can be passed to functions as a configuration to be utilized over and over for reuse of the bus configuration.

Other features that have been extended in NServiceBus in the newer versions also include the addition of NSB packages. One such package is `NServiceBus.NLog`, which is specific to using NLog in NSB version 5.0 and above. This is used to extend the logging capabilities of NSB to log further information as it applies to endpoints, messages, and services. Working with the `NServiceBus.NLog` package has been detailed in `http://docs.particular.net/nservicebus/logging-in-nservicebus`. We will demonstrate the code's use further in this chapter.

The saga workflow

The purpose of the saga is to provide workflow. It provides the persistence of the saga data so that a message can return to the saga and have a point of reference to the original messages.

The source code in this section is in the *CreditCardApproval – v5* section. It was compiled in Visual Studio 2012 using NSB Version 5.0. The projects will appear as the following:

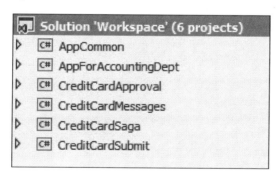

Here we have the following projects:

- `AppCommon`: This contains the ViewModel and Context for the Windows forms.
- `AppForAccountingDept`: This receives a purchase order number related to the approval of the credit card.
- `CreditCardApproval`: This approves the credit card.
- `CreditCardSubmit`: This submits the credit card for approval.

- `CreditCardMessages`: This contains the common messages for the projects.
- `CreditCardSaga`: This is the credit card saga that directs the workflow and message routing by using message handling and saga data.

The flow of the messages, which does all the work (thus, workflow), will start with the `CreditCardSubmit` project. This form will submit a credit card to the saga, which will save its state and forward it to the `CreditCardApproval` project. The `CreditCardApproval` form will either approve or deny the card and send the response back to the saga. The saga will retrieve the previous message and pass it back to the `CreditCardSubmit` form, and if approved, will send it to the `AppForAccountingDept` project. The message flow will appear as follows:

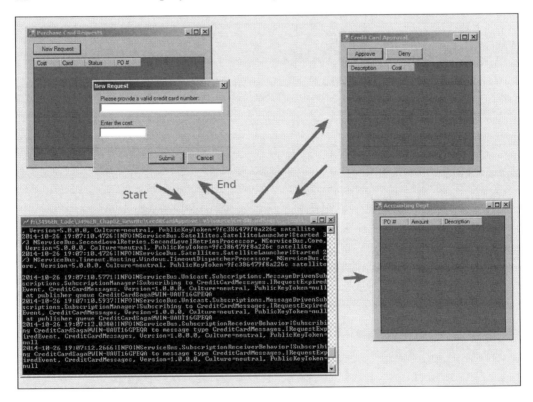

This is a very simplistic workflow as it keeps track of whether a credit card is approved or not. By making the saga data persistent, it allows the persistence of the saga data state to route the messages and manage the message information at a single point. The saga data will contain the required ID, originator, and `OriginalMessageId`, to ensure a response to the original client. These fields ensure that the return is always available. We see these fields in the following diagram:

Chapter 2

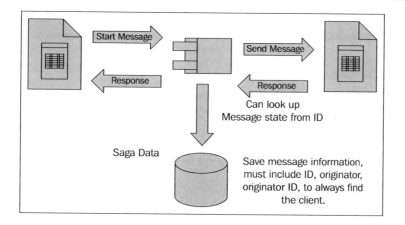

The saga data is stored by using the IContainSagaData interface. Even if the three fields are not explicitly defined, the IContainSagaData interface will define them in storage. Do not modify these fields as NServiceBus uses them. The primary key is another field that is used in all sagas to look up the saga data from the message. In this case, we use the RequestId that will always contain a unique value from the [Unique] annotation. We can see this in the following code:

```
public class CreditCardRequestData : IContainSagaData
{
    /***
     * Gets/sets the Id of the process. Do NOT generate this value in your code.
       The value of the Id will be generated automatically to provide the
       best performance for saving in a database.
     * ***/
    public Guid Id { get; set; }  // Required
    /***
     * Contains the return address of the endpoint that caused the process to run.
     * ***/
    public string Originator { get; set; }  //Required
    /***
     * Contains the Id of the message which caused the saga to start.
       This is needed so that when we reply to the Originator, any
       registered callbacks will be fired correctly.
     * ***/
    public string OriginalMessageId { get; set; }  //Required

    [Unique]
    public Guid RequestId { get; set; }  // Unique ID to lookup Request message
    public string Card { get; set; }
    public decimal Cost { get; set; }
    public bool RequiresApproval { get; set; }
    public bool Approved { get; set; }
}
```

[43]

This saga data will be long-lived transactional data that will live in the database until the saga does a `MarkAsComplete()` function on it to designate that it is no longer required. If we use `InMemoryPersistence`, the data will only live in memory during the lifetime of the application, and not persist in a database. In version 5.0, all the persistence can be set with a single function call, `busConfiguration.UsePersistence<InMemoryPersistence>();`. It can still be pieced out for `Saga`, `Timeout`, `Gateways`, and `Subscriptions` by using extended references in this function as `busConfiguration.UsePersistence<InMemoryPersistence>().For(Storage.Sagas);`. These storage types can be found in the `NServiceBus.Persistence` namespace as seen in the following code:

```
namespace NServiceBus.Persistence
{
    // Summary:
    //     The storage needs of NServiceBus
    public enum Storage
    {
        // Summary:
        //     Storage for timeouts
        Timeouts = 1,
        //
        // Summary:
        //     Storage for subscriptions
        Subscriptions = 2,
        //
        // Summary:
        //     Storage for sagas
        Sagas = 3,
        //
        // Summary:
        //     Storage for gateway deduplication
        GatewayDeduplication = 4,
        //
        // Summary:
        //     Storage for the outbox
        Outbox = 5,
    }
}
```

Another way to use the `MarkAsComplete()` function is by way of the timeout in the saga service.

The saga service can provide timeouts to give the message a limited life. This is a very useful function for many purposes; for example, when there is an error and the message is taking forever, for instance, in processing a credit card, and the bank received the message but never provided a response.

Chapter 2

The timeout could also be used to schedule daily tasks for the saga to complete in its cleanup and maintenance pieces. For instance, we can send a daily e-mail to operations reporting how many messages the saga processed for the day and if it had any messages reported in the error queue. Here, we give the credit card approver 60 seconds to approve. The following screenshot displays the code for this:

```
public void Handle(SubmitRequestCommand message)
{
    logger.Info("--------CreditCardSaga Handle-------" + message);

    // We give the Approver 60 seconds to respond
    RequestTimeout<TimeoutMessage>(TimeSpan.FromSeconds(60));
```

The saga's purpose is to manage the workflow of the messages by the state of the saga data. The saga will be started by a message, or many messages, defined in IAmStartedByMessages<>. Subsequent messages are defined in IHandleMessages< >. The timeout has a special message handler called IHandleTimeouts<>. We can observe this in the following class diagram:

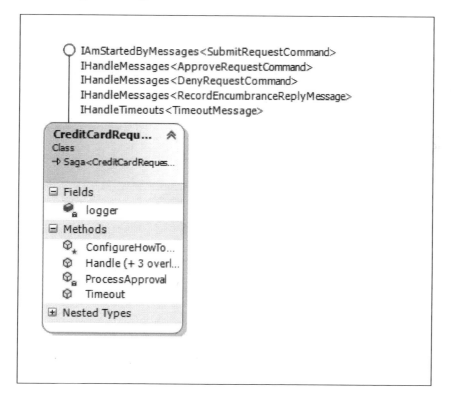

[45]

In order to save the saga data, the data needs to be mapped to a message. NSB will do most of the mapping, but it requires the definition of the message ID that will be used to map the data to the message. In our saga data, we are using the `RequestId` instance as the message's data primary key. In version 4.0, we mapped this functionality to use the `RequestId` instance to map the messages `ApproveRequestCommand` and `DenyRequestCommand`, as shown in the following code, using `ConfigureHowToFindSaga()`:

```
public override void ConfigureHowToFindSaga()
{
    ConfigureMapping<ApproveRequestCommand>(x => x.RequestId).ToSaga(x => x.RequestId);
    ConfigureMapping<DenyRequestCommand>(x => x.RequestId).ToSaga(x => x.RequestId);
}
```

In version 5.0, we mapped this functionality to use the `RequestId` instance to map the messages `ApproveRequestCommand` and `DenyRequestCommand`, as shown in the following code, using `ConfigureHowToFindSaga(SagaPropertyMapper< > mapper)`:

```
protected override void ConfigureHowToFindSaga(SagaPropertyMapper<CreditCardRequestData> mapper)
{
    mapper.ConfigureMapping<ApproveRequestCommand>(x => x.RequestId).ToSaga(x => x.RequestId);
    mapper.ConfigureMapping<DenyRequestCommand>(x => x.RequestId).ToSaga(x => x.RequestId);
}
```

These messages are handled by message handlers, such as the `Handle(DenyRequestCommand message)`. The saga data will pull the record matching the message's `RequestId` from NSB. NSB just needs the mapping definition, and it handles the record that is maintaining persistence and the mapping, as shown in the following code:

```
public void Handle(DenyRequestCommand message)
{
    logger.Info("---------CreditCardSaga Handle---------" + message);
    var reply = new SubmitRequestReplyMessage
    {
        RequestId = Data.RequestId,
        Approved = false
    };

    ReplyToOriginator(reply);
    MarkAsComplete();
}
```

same

In this function, the `ReplyToOriginator(message)` function will send the message to the original instance of the client. The `MarkAsComplete()` function will tell the persistence that the saga's data record matching the message's `RequestId` instance can now be deleted.

For NLog, we set the NSB NLog as follows:

```csharp
public void Customize(BusConfiguration busConfiguration)
{
    // NLog the Bus
    NServiceBus.Logging.LogManager.Use<NLogFactory>();

    logger.Info("--------Credit Card Saga Configure--------");
    busConfiguration.UsePersistence<InMemoryPersistence>();

    busConfiguration.UseTransport<MsmqTransport>();

    logger.Info("--------Credit Card Saga Running--------");

}
```

The configuration has to be set for the logging levels and the location where the log files will be saved in the `App.config` file, as shown here:

```xml
<nlog xmlns="http://www.nlog-project.org/schemas/NLog.xsd" xmlns:xsi="http://www.w3.org/2001/XMLSchema-instance">
    <targets>
        <target name="logfile" xsi:type="File" fileName="CreditCardSaga_${shortdate}.log" layout="${longdate} ${level} ${message}" />
        <target name="console" xsi:type="Console" />
        <target xsi:type="EventLog" name="event" layout="${message}" source="MyProgram" eventId="555" log="Application" />
    </targets>
    <rules>
        <logger name="*" minLevel="Error" writeTo="event" />
        <logger name="*" minLevel="Info" writeTo="console" />
        <logger name="*" minLevel="Trace" writeTo="logfile" />
    </rules>
</nlog>
```

This will log the lower-level messages in NSB, as we can see in the following screenshot:

```
CreditCardSaga_2014-10-27 - Notepad
File  Edit  Format  View  Help
2014-10-27 00:28:33.8492 Info --------Credit Card Saga Configure--------
2014-10-27 00:28:33.8805 Info --------Credit Card Saga Running--------
2014-10-27 00:28:34.2555 Info Activating persistence 'InMemoryPersistence' to provide storage for 'Sagas' storage.
2014-10-27 00:28:34.2711 Info Activating persistence 'InMemoryPersistence' to provide storage for 'Timeouts' storage.
2014-10-27 00:28:34.2711 Info Activating persistence 'InMemoryPersistence' to provide storage for 'Subscriptions' storage.
2014-10-27 00:28:34.2711 Info Activating persistence 'InMemoryPersistence' to provide storage for 'Outbox' storage.
2014-10-27 00:28:34.2711 Info Activating persistence 'InMemoryPersistence' to provide storage for 'GatewayDeduplication' storage.
2014-10-27 00:28:34.4117 Debug Error queue retrieved from <MessageForwardingInCaseOfFaultConfig> element in config file.
```

NServiceBus Saga Architecture

Message flow

The originating client that starts the process will be `CreditCardSubmit`. This client could easily be a website or another means of starting the messaging process. The `SubmitRequestCommand` project is the starting message for the saga and the message is sent from the `CreditCardSubmit` project. It is defined in the `App.config` file of the `CreditCardSubmit` project to send to the saga in the following code:

```
<UnicastBusConfig ForwardReceivedMessagesTo="MyAudits">
  <MessageEndpointMappings>
    <add Endpoint="CreditCardSaga" Messages="CreditCardMessages.SubmitRequestCommand, CreditCardMessages" />
    <add Endpoint="CreditCardSaga" Messages="CreditCardMessages.IRequestExpiredEvent, CreditCardMessages" />
  </MessageEndpointMappings>
</UnicastBusConfig>
```

We can view the messages being sent in MSMQ, but if the saga was running, the messages would process too fast to observe. We can see the message in the saga without running all the services to take a snapshot in time, as we see in the following screenshot

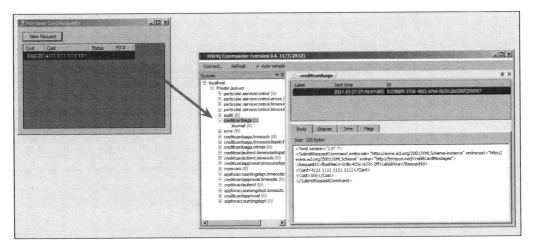

Once the saga starts, it handles the `SubmitRequestMessage` message. The saga will populate the data with defaults and message information. Notice that `RequestId` is set in the saga data to match the message's `RequestId` instance. This is to map the saga data to the other messages that the saga handles, but in order for this `RequestId` instance to match, this key must be set in each message that the saga needs to map. The code for this is shown in the following screenshot:

Chapter 2

```
public void Handle(SubmitRequestCommand message)
{
    logger.Info("--------CreditCardSaga Handle-------" + message);

    // We give the Approver 60 seconds to respond
    RequestTimeout<TimeoutMessage>(TimeSpan.FromSeconds(60));

    Data.RequestId = message.RequestId;
    Data.Card = message.Card;
    Data.Cost = message.Cost;
    Data.RequiresApproval = true;
    Data.Approved = false;

    ProcessApproval();
}
```

After we save the saga data, we will create a `SolicitApprovalFromLevel1Command` message and send it from the bus, as shown in the following code:

```
private void ProcessApproval()
{
    logger.Info("--------CreditCardSaga Process Approval-------");
    if (Data.RequiresApproval && !Data.Approved)
    {
        logger.Info("--------SolicitApprovalFromLevel1Command-------");
        var command = new SolicitApprovalFromLevel1Command
        {
            RequestId = Data.RequestId,
            Card = Data.Card,
            Cost = Data.Cost
        };

        Bus.Send(command);
        return;
    }
}
```

Downloading the example code

You can download the example code files for all Packt books you have purchased from your account at http://www.packtpub.com. If you purchased this book elsewhere, you can visit http://www.packtpub.com/support and register to have the files e-mailed directly to you.

NServiceBus Saga Architecture

The bus knows which MSMQ to place from the `App.config` file onto the `CreditCardApproval` queue. The `App.config` file of the saga will show the queue names that it will send to, based on the message type. We see that the `CreditCardApproval` queue will get the `SolicitApprovalFromLevel1Command` message as follows:

```xml
<UnicastBusConfig ForwardReceivedMessagesTo="audit">
    <MessageEndpointMappings>
        <add Endpoint="CreditCardApproval" Messages="CreditCardMessages.SolicitApprovalFromLevel1Command, CreditCardMessages" />
        <add Endpoint="AppForAccountingDept" Messages="CreditCardMessages.RecordEncumbranceCommand, CreditCardMessages" />
    </MessageEndpointMappings>
</UnicastBusConfig>
```

We can also observe the MSMQ queue with a message, which helps in tracking visual reporting on the message flow as it happens:

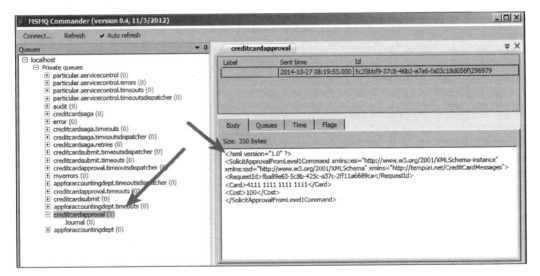

Once the message is in the `CreditCardApproval` queue, the `CreditCardApproval` program can handle the message. The program will define what types of messages it will handle, and in this case, the NSB message handler will populate the Windows forms to allow us to approve or deny the card. Instead of a simple Windows form manual approval, the approval could have easily been a credit card approval sent to a bank, made through an internal business process, sent to a mainframe, or through any other scenario. NSB is built to develop ESBs in C# and interface into many different types of systems. So, instead of migrating an entire system, it is easier to use legacy bits and pieces, until migration is necessary. Here's the message handler that loads the Windows forms:

Chapter 2

```csharp
public class SolicitApprovalFromLevel1CommandHandler : IHandleMessages<SolicitApprovalFromLevel1Command>
{
    public Context<ItemViewModel> Context { get; set; }

    public void Handle(SolicitApprovalFromLevel1Command message)
    {
        Context.MarshalToUiThread(() => HandleOnUiThread(message));
    }

    private void HandleOnUiThread(SolicitApprovalFromLevel1Command message)
    {
        var item = new ItemViewModel
        {
            RequestId = message.RequestId,
            Card = message.Card,
            Cost = message.Cost
        };

        Context.Items.Add(item);
    }
}
```

Also note that in a production system, credit cards would be encrypted. As we have discussed in the previous chapter, NSB easily handles the encryption of data and messages. This will load the selection for approval or denial as follows:

[51]

NServiceBus Saga Architecture

If we click on the **Approve** button, an `ApprovalRequestCommand` message will be sent back to the saga. If we click on the **Deny** button, a `DenyRequestCommand` message will be sent back to the saga. This is shown in the following screenshot:

```csharp
private void ButtonApproveClick(object sender, EventArgs e)
{
    var item = bindingSource.Current as ItemViewModel;
    if (item == null)
    {
        MessageBox.Show("Nothing selected to approve.");
        return;
    }

    Context.Items.Remove(item);
    Bus.Send(new ApproveRequestCommand { RequestId = item.RequestId, Approver = Approver.Level1 });
}

private void ButtonDenyClick(object sender, EventArgs e)
{
    var item = bindingSource.Current as ItemViewModel;
    if (item == null)
    {
        MessageBox.Show("Nothing selected to deny.");
        return;
    }

    Context.Items.Remove(item);
    Bus.Send(new DenyRequestCommand { RequestId = item.RequestId });
}
```

The messages will be placed back on the saga queue for processing using the message handler. Notice that the messages have the same `RequestId` instance as the original message so that the saga data can be retrieved from the previous message. This is shown in the following screenshot:

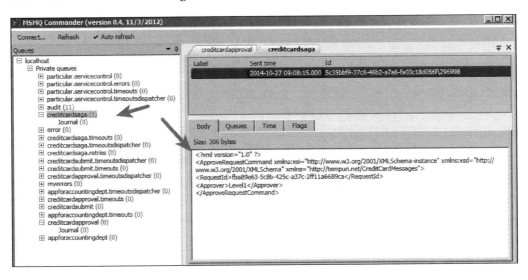

The saga's message handlers will process the approval or denial message and reply back to the originating client, which is the `CreditCardSubmit` application, as shown here:

```csharp
public void Handle(ApproveRequestCommand message)
{
    logger.Info("--------CreditCardSaga Handle-------" + message);
    if (message.Approver == Approver.Level1)
    {
        Data.Approved = true;
    }

    ProcessApproval();
}

public void Handle(DenyRequestCommand message)
{
    logger.Info("--------CreditCardSaga Handle-------" + message);
    var reply = new SubmitRequestReplyMessage
    {
        RequestId = Data.RequestId,
        Approved = false
    };

    ReplyToOriginator(reply);
    MarkAsComplete();
}
```

If approved, the saga will send the message to the accounting department to generate a purchase order, and respond to the saga with the Purchase Order, as shown here:

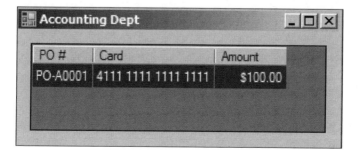

NServiceBus Saga Architecture

The saga will read and handle the message from the accounting department of the type, `RecordEncumbranceReplyMessage`. The saga will respond to the original client, `CreditCardSubmit` so that it can be updated with the Purchase Order number and status. We can see this in the saga message handler code in the following screenshot:

```
public void Handle(RecordEncumbranceReplyMessage message)
{
    logger.Info("--------CreditCardSaga Handle-------" + message);
    var reply = new SubmitRequestReplyMessage
    {
        RequestId = Data.RequestId,
        Approved = true,
        PurchaseOrderNumber = message.PurchaseOrderNumber
    };

    ReplyToOriginator(reply);
    MarkAsComplete();
}
```

The originating client will receive the Purchase Order, via the saga from the accounting department application, as shown here:

After the original requester receives the Purchase Order, the message is complete. For denial and errors, the message flow will be similar.

One may ask why there is so much work in a message flow. NSB is doing almost all the work for the message flow and mapping. Many organizations save data for the purpose of handling errors. The saga has a lot of basic error handling mechanisms for handling errors throughout messages, endpoints, and services. We have already put in code for timeouts so that the messages occur within a specific period of time.

Sagas assist in providing various kinds of reporting for error handling. We know for sure that the message is always durable in MSMQ; the machine could be restarted and the message will still reside in MSMQ if it is supposed to persist. We know that the saga data would normally reside in the database. We only use InMemory persistence during debugging. What we have shown is that we can study the minute details of how message flows, endpoints, and services react with NSB, and how we can use the ESB to provide feedback on how all the pieces work as a unit to bring end-to-end workflow through the use of saga. We have also observed that in any message, endpoint, and service, we can extend feedback to the originating client of any condition, such as network or server condition, to recover the original intent of the message and recover the functionality. NSB has shown to be a robust and durable framework for any message or data that we pass through it. We can truly see its value in end-to-end workflows.

Deployment

All applications are executable, except the saga, which we developed as a DLL because of the use of `NServiceBus.Host.exe`. This is a program and package from NSB to provide hosted deployment from NSB, in which we can install the DLL as a managed service and add endpoints and many other features that NSB hosts bring to the table. When we debug the program, we start the saga DLL using the `NServiceBus.Host.exe` program. See the following project properties:

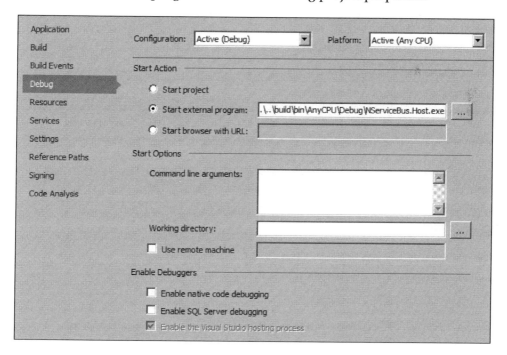

NServiceBus Saga Architecture

To run the program as a console application, we could simply use `NServiceBus.Host.exe CreditCardSaga.dll` as shown in the following screenshot:

```
Administrator: C:\Windows\system32\cmd.exe - NServiceBus.Host.exe  CreditCardSaga.dll
ug>NServiceBus.Host.exe   CreditCardSaga.dll
2014-10-27 11:06:00.220 INFO  DefaultFactory Logging to 'F:\3496EN_Code\3496EN_C
hap02_Rewrite\CreditCardApprove - v5\build\bin\AnyCPU\Debug\' with level Info
2014-10-27 11:06:00.9709|INFO|CreditCardSaga.EndpointConfig|----------Credit Card
Saga Configure----------
2014-10-27 11:06:01.4240|INFO|CreditCardSaga.EndpointConfig|----------Credit Card
Saga Running----------
2014-10-27 11:06:01.486 INFO  NServiceBus.Hosting.Profiles.ProfileManager Activa
ting profile: NServiceBus.Production, NServiceBus.Host, Version=5.0.0.0, Culture
=neutral, PublicKeyToken=9fc386479f8a226c
2014-10-27 11:06:02.1896|INFO|NServiceBus.Persistence.PersistenceStartup|Activat
ing persistence 'InMemoryPersistence' to provide storage for 'Sagas' storage.
2014-10-27 11:06:02.2052|INFO|NServiceBus.Persistence.PersistenceStartup|Activat
ing persistence 'InMemoryPersistence' to provide storage for 'Timeouts' storage.
2014-10-27 11:06:02.2209|INFO|NServiceBus.Persistence.PersistenceStartup|Activat
ing persistence 'InMemoryPersistence' to provide storage for 'Subscriptions' sto
rage.
2014-10-27 11:06:02.2365|INFO|NServiceBus.Persistence.PersistenceStartup|Activat
ing persistence 'InMemoryPersistence' to provide storage for 'Outbox' storage.
2014-10-27 11:06:02.2521|INFO|NServiceBus.Persistence.PersistenceStartup|Activat
ing persistence 'InMemoryPersistence' to provide storage for 'GatewayDeduplicati
on' storage.
2014-10-27 11:06:06.7209|INFO|NServiceBus.Licensing.LicenseManager|UpgradeProtec
tionExpiration: 10/4/2015 12:00:00 AM
```

We could also use the `NServiceBus.Host.exe\install` to install the program as an NSB-managed service. For more details on NSB hosting, please visit http://docs.particular.net/nservicebus/the-nservicebus-host.
Next, let's look at the insight.

ServiceInsight

We covered the topic of architecture using some basic tools to view the MSMQ as the messages were flowing. NSB offers ServiceInsight that can offer much more insight to the flow and execution of messages, endpoints, and services. In order to view the messages, we have to install ServiceInsight on the machine along with ServiceControl, and add the ServiceControl debugging and saga packages to the saga project. We will need to add service control to the other projects if we wish to view them as well. In this instance, we will just add service control debug and saga to `CreditCardSaga`. We will add references through NuGet to appear as in the following screenshot:

Chapter 2

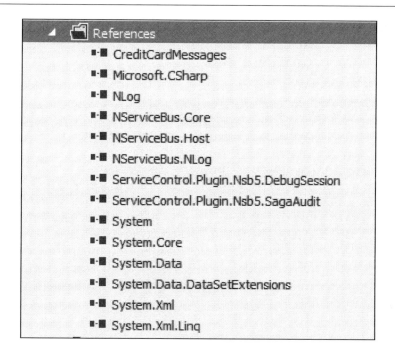

After adding service control to the project, which we will call CreditCardApprove - v5 - ServiceInsight, we can get detailed information on the saga and its messages through ServiceInsight, which will look similar to what is seen in the following screenshot:

Summary

We have discussed moving sagas from version 4.0 to version 5.0. Version 5.0 has many more tools and packages that are geared towards Windows Azure, Visual Studio 2012, and .NET 4.5. This enables greater compatibility with Azure development and testing, as well as using Azure SDK tools, and the many tools that Visual Studio 2012 provides. This is definitely a positive sign in the evolution of a saga with Windows cloud computing.

We discussed a credit card approval sample using a saga service and saga data. This example could be used in multiple scenarios: from using the WCF service of a bank to deduct from a credit card to ordering from a catalog. The messages, endpoints, and services are managed by NSB as it maps the state of the messages and ensures their durability. The message flow is an important concept as messages can be sent to different business divisions for their business applications, or handled as separate processes, just as we used an accounting department application. The messaging is viewed and reported through multiple means. One of those means includes NServiceBus tools, such as ServiceInsight, which will be discussed in further detail in the next chapter.

There are also multiple ways to deploy the application, depending on whether the `NServiceBus.Host` executable is used or not. The reason for the use of the `NServiceBus.Host` executable and framework is to have NSB manage the services, endpoints, and messages more as a managed service. NSB is a complete end-to-end ESB that provides many benefits beyond just an ESB in creating, developing, testing, and deploying applications in a Windows C# environment.

3
The Particular Service Platform

In this chapter, we will be focusing on the Particular Service Platform, which includes ServicePulse, ServiceControl, ServiceInsight, and ServiceMatrix. We can use ServicePulse to get a pulse of **NServiceBus (NSB)** endpoints, messages, sagas, and services. This is more of a production monitoring product, for operations to check for running NSB components.

ServiceInsight is the product to use to get into the details of NSB endpoints, messages, sagas, and services, to drill down into issues, and to verify proper operations in detail. ServiceMatrix is the graphical developer interface that extends into a Visual Studio canvas for code generation of NSB endpoints, services, and messages.

In this chapter, we will cover:

- ServicePulse
- ServiceControl
- ServiceInsight
- ServiceMatrix
- Publish-subscribe through ServiceMatrix
- Sagas through ServiceMatrix
- Introducing custom checks for ServicePulse

Introducing NSB components

There are many tools that can be licensed from `http://particular.net/` which aid in developing and monitoring NSB components. NSB components include endpoints, messages, and services. Components include any of the configured items on the IBus.

ServiceMatrix is an extension plugin that can be installed into Visual Studio 2012. This plugin creates an NSB canvas that creates various NSB components through wizards, which will generate C# code to be used for the ESB. ServiceMatrix simplifies and standardizes the development lifecycle by generating code to match diagrams designed in the NSB Canvas, and helps developers create NSB-standardized code from the NSB Visual Studio wizard. NSB wizards assist in learning how different pieces make up an NSB application from end to end. ServiceMatrix also creates skeleton pieces and associate generated C# code to run an end-to-end NSB solution with less effort from the developer.

ServiceControl is a controller that is a management extension to NSB. Its code can be downloaded from `https://github.com/Particular/ServiceControl`. ServiceControl is a Windows service that collects endpoints, messages, and service information, and stores it in a local RavenDB. This is done so that other products such as ServicePulse and ServiceInsight can access the information to report on the heartbeats of services and give detailed information regarding messages. ServicePulse checks for heartbeats on NSB applications with the heartbeat plugin, and provides custom checks as well. It has a dashboard that shows which NSB applications are up and running for production. ServiceInsight provides details of messages, endpoints, and services. It shows a detailed graphical flow of messages and properties of messages that includes endpoints and service information, timeouts, payloads, and more. ServiceInsight allows a deep-dive into messaging. These tools can be found from the download page, `http://particular.net/downloads`. While it is possible to monitor, maintain, and build NSB applications without these tools, there is a lot of work and effort that has gone into these tools to help the developer understand, build, and maintain NSB applications, to get a product that is better built and is made available to the market faster.

We will walk through setting up ServicePulse, then ServiceInsight, and then build some applications with ServiceMatrix to be examined by both, ServicePulse and ServiceInsight. ServiceControl is an install that both ServicePulse and ServiceInsight will need to use to gather its information. We will start with ServicePulse to show us a dashboard as it is the most basic of these tools. To run any of the applications in this chapter, ServiceControl, ServicePulse, ServiceMatirx, and ServiceInsight must be installed. You can download these by going to `http://particular.net/downloads` and clicking on the **Download Now** button to select all the products through a Platform Installer, which will work on Windows 8.1 and other desktop machines. For some servers such as Windows Server 2008, this button may not work, and individual downloads from this page may have to be installed separately. This chapter assumes that all these products are installed.

Understanding ServicePulse and its function

ServicePulse is an operational monitoring tool for applications in NSB. It has three main functions: monitoring heartbeats, monitoring errors, and retrying extensibility for custom checks.

We can get a dashboard of failed messages, endpoint heartbeats, successful messages, and custom checks as shown in the following screenshot:

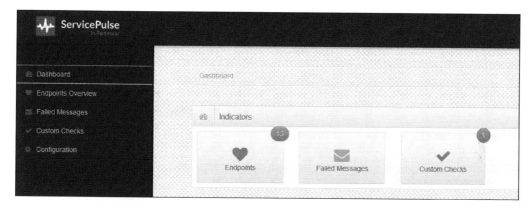

Besides the dashboard, we can get endpoint overviews, failed messages, custom checks, and configurations as shown in the following screenshot:

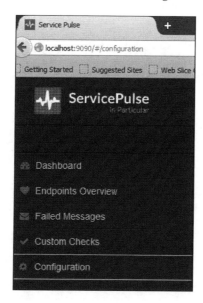

ServicePulse is an important tool as it can tell us which endpoints are running. In order for ServicePulse to be running at all, ServiceControl will have to be installed. It is important to have ServiceControl operational to run ServicePulse and ServiceInsight.

Understanding ServiceControl and its function

As mentioned before, for ServicePulse to work, ServiceControl has to be installed. ServiceControl can be downloaded from `http://particular.net/downloads`.

ServiceControl is a service that is used for auditing and monitoring NSB endpoints, messages, and applications, and saves it in a local instance of RavenDB. It also grants access to these gathered audited messages forwarded by NSB endpoints through an exposed **JavaScript Object Notation (JSON)** HTTP API, which provides data and functionality services for ServiceInsight and ServicePulse. Since RavenDB uses JSON through an HTTP API by default, ServiceControl acts as a collection service to gather the information that is important for NSB tools. ServiceControl configuration and troubleshooting instructions can be found at `http://docs.particular.net/servicecontrol/`.

When installing ServiceControl as a separate package, a window similar to the following screenshot will appear on your screen:

ServiceControl will use `http://localhost:33333/api` by default. The ports are configurable as we see in the previous screenshot. ServiceControl supports other queuing types, such as SQL Server queues, Azure, and RabbitMQ. You will find instructions on this at `http://docs.particular.net/servicecontrol/multi-transport-support`.

ServiceControl normally runs through the URL at `http://localhost:33333/api`. If the ServiceControl screen does not come up correctly, you may want to check if the Particular ServiceControl Windows service has started. ServiceInsight and ServicePulse will be looking to read the endpoint information from this port. We can see this API start in this screenshot:

Notice that this is the programmatic API for monitoring the IBus. This is your natural hook point into monitoring systems and other integrations. You can develop your own operation automation that reacts to the state of the bus. This is a JSON messaging system through HTTP where you can walk down the different endpoints, messages, and further information gathered by ServiceControl. More on (JSON) can be found at `http://www.json.org`.

What does this mean? We can start walking down the JSON data at this starting point. We can view all the messages by calling `http://localhost:33333/api/messages/` and all the endpoints by viewing `http://localhost:33333/api/endpoints/`. We can then start creating our own GUI in C# by calling the JSON API, and viewing a particular endpoint by passing in the endpoint name.

ServicePulse will know how to call this starting point of the JSON API by its setting when ServicePulse is installed. Please see the following screenshot to see where we set the ServiceControl API instance:

Let's look at a simple example. We can start with the publish-subscribe MSMQ example from `https://github.com/Particular/NServiceBus.Msmq.Samples/tree/master/PubSub`.

We will need to add ServiceControl plugins through NuGet to generate ServiceControl endpoints for monitoring purposes, otherwise there will be nothing to monitor ServiceControl. The following plugins are currently available:

- `ServiceControl.Plugin.DebugSession`: This is found at `https://www.nuget.org/packages/ServiceControl.Plugin.Nsb4.DebugSession/` for version 4.x and `https://www.nuget.org/packages/ServiceControl.Plugin.Nsb5.DebugSession/` for version 5.x. When deployed, the debug session plugin adds a specified debug session identifier to the header of each message sent by the endpoint. This allows messages sent by debugging or a test run within Visual Studio to be correlated, filtered, and highlighted within ServiceInsight.

Chapter 3

- `ServiceControl.Plugin.CustomChecks`: This is found at https://www.nuget.org/packages/ServiceControl.Plugin.Nsb4.CustomChecks/ and for version 4.x and at https://www.nuget.org/packages/ServiceControl.Plugin.Nsb5.CustomChecks/ for version 5.x. The result of a custom check is either a success or a failure (with a detailed description defined by the developer). This result is sent as a message to the ServiceControl queue.

- `ServiceControl.Plugin.Heartbeat`: This is found at https://www.nuget.org/packages/ServiceControl.Plugin.Heartbeat. The heartbeat plugin sends heartbeat messages from the endpoint to the ServiceControl queue. These messages are sent every 10 seconds by default.

`ServiceControl.Plugin.SagaAudit`: This is found at https://www.nuget.org/packages/ServiceControl.Plugin.Nsb4.Heartbeat/ for version 4.x and at https://www.nuget.org/packages/ServiceControl.Plugin.Nsb5.Heartbeat/ for version 5.x. The Saga Audit plugin collects the activity information of a saga runtime. This information enables the display of detailed saga data, behaviors, and the current status in ServiceInsight Saga View. The plugin sends the relevant saga state information as messages to the ServiceControl queue whenever a saga state changes.

To add the plugins, perform the following steps:

1. We will add the ServiceControl plugins for heartbeats and custom checks through NuGet:

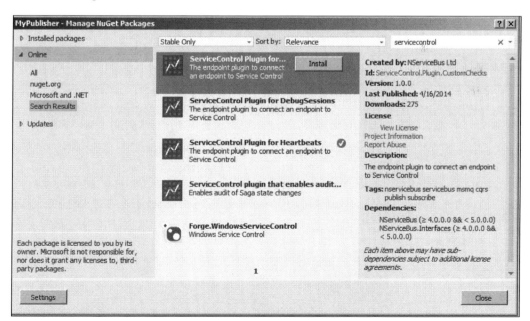

2. Then, we can check heartbeats in ServicePulse to validate that the applications are giving heartbeats that indicate availability. We monitor ServicePulse through the URL, `http://localhost:9090`. This can be seen in the following screenshot for ServicePlus

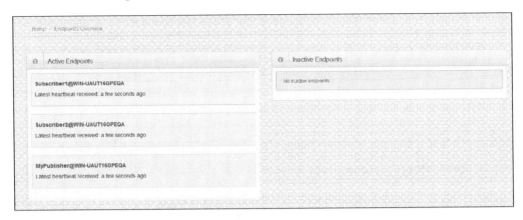

3. If there are issues with the services, always check that ServiceControl and ServicePulse are running. The following screenshot shows these running:

Understanding ServiceInsight and its function

ServiceInsight provides detailed insights into a specific message. It provides detail flow, timing, and error handling and the ability to retry the message, sort the message, look at its header, look at its sagas, copy the header, copy the message, and more.

We will explore this more with the ServiceMatrix examples that we will be building, but we need to familiarize ourselves with the functions of ServiceInsight. You may opt to not use some of these tools in your development, but the purpose of this chapter is to discuss these tools.

We have the Endpoint Explorer, which gives us details about a message, and a **Message Properties** window to drill down into the details of the message. We also have a **Flow Diagram** window to give us a graphical overview of the message and endpoint. Please see the following screenshot to view the **Messages**, **Message Properties**, and a **Saga** flow view:

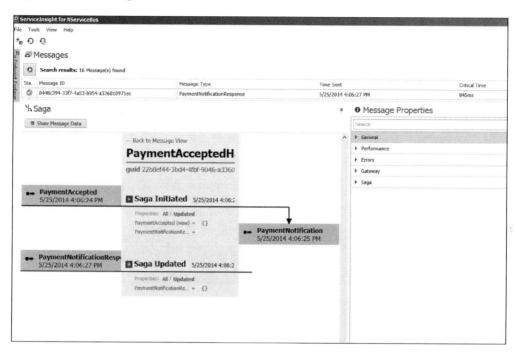

This flow view is very important. Here is the graphical picture of what the NSB application thinks it is behaving like. ServiceControl puts together different message properties and endpoint information from ServiceControl to derive the flow view.

The **Endpoint Explorer** window gives a list of the available endpoints that have been captured in ServiceControl. This list can be used to filter all of the available messages so that you may view just the messages on an endpoint. The following is an example of an Endpoint Explorer tree:

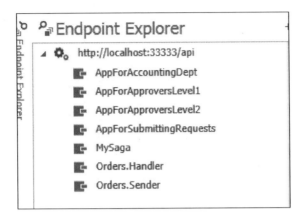

However, it is not a requirement to use ServiceMatrix to build NSB components, as an NSB solution can be created from programming C# code. ServiceMatrix will generate the skeleton pieces of C# code and ensure that the programmer uses the generated code in a standard NSB format that will match the NSB tools in Visual Studio.

A step-by-step guide to use ServiceMatrix can be found at `http://docs.particular.net/servicematrix/getting-started-with-servicematrix-2.0`. A guide to using code without ServiceMatrix can be found at `http://particular.net/articles/NServiceBus-Step-by-Step-Guide`.

We will walk through our own solution from start to finish in order to create a solution for a Payment Engine in a request-response message flow, in a directory called `PaymentEngine - Start`. This solution will be the end result of this section and it will contain ServiceControl plugins to monitor it through ServicePulse and ServiceInsight.

Through in the next sections of this chapter on ServiceMatrix, we will take this request-response solution in the `PaymentEngine - Start` directory and extend it to a publish-subscribe message flow, with the addition of sagas. The final result will be that publish-subscribe and sagas will be in the `PaymentEngine - Sagas` directory. In Windows 8.1, you may have to run Visual Studio 2012 as the administrator. When running Visual Studio 2012 on Windows Server 2008 and Windows Server 2012, you may not have to run Visual Studio 2012 as the administrator.

Creating a ServiceMatrix solution

We will install ServiceMatrix in Visual Studio by navigating to **Tools | Extensions and Updates**. You will be presented with the following screen:

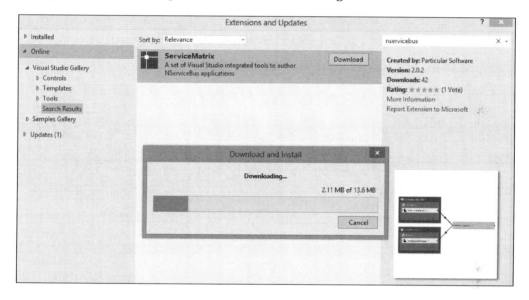

The Particular Service Platform

We can create an NSB project by navigating to **Files** | **New** | **Project**. Here, we will create a Payment Engine example. Let's start by creating a ServiceMatrix solution called `PaymentEngine` in Visual Studio 2012, as in the following screenshot. Please note that it starts out as a solution type of the NSB System.

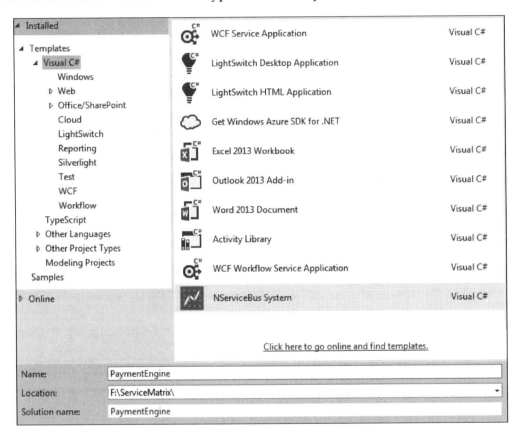

Normally, there are three different areas for the standard development environment. There is **Solution Builder** on the left, **NServiceBus Canvas** in the center, and **Solution Explorer** on the right. This is shown in the following screenshot:

Chapter 3

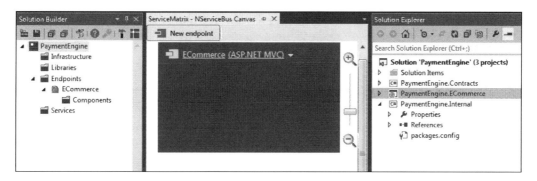

We will create another endpoint called `PaymentProcessing` that will be an NSB Host program. The NSB host streamlines service development and deployment, allows you to change technologies without code, and is administrator-friendly when setting permissions and accounts. Visit `http://docs.particular.net/nservicebus/the-nservicebus-host` for more information.

We can select the Solution Builder or click on **New endpoint** in **NServiceBus Canvas** and type in the name. The ServiceMatrix will then generate the code to create the new endpoint, as shown in the following screenshot:

The Solution Builder contains four main sections:

- **Infrastructure:** This is used to create and manage NSB authentication and auditing
- **Libraries**: This is used to create and manage NSB reusable libraries
- **Endpoints**: This is used to create and manage NServiceBus endpoints
- **Services**: This is used to create and manage NServiceBus services

[71]

The Particular Service Platform

By right-clicking on the elements of these sections, we can add or change properties, as shown in the following screenshot:

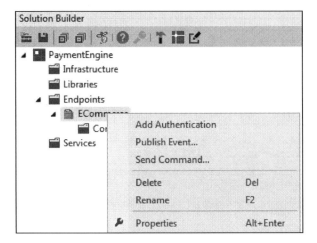

We can accomplish similar tasks in **NServiceBus Canvas**. The difference is that this is a visual graph showing the flow instead of a tree directory hierarchy. Right-click on the graphical information as shown in the following screenshot:

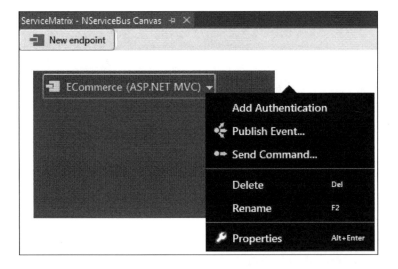

The **Solutions Explorer** pane will display the resultant generated code. Some of the code will be stubs that are created to add more detail during development. An event can be created through **Publish Event...** and a command message can be created through **Send Command...**. We can create a send command message. We will name the service `Payments` for the `SubmitPayment` command message as in the following screenshot:

Chapter 3

The **Contracts** section will contain NSB events, and the **Internal** section will contain NSB commands. Notice that a `SubmitOrder.cs` file was created when we created the `SubmitPayment` command. Here is where we will find the C# code file:

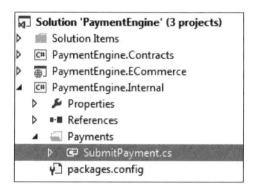

We can see the code that would normally contain your command message, at this point, is but a code stub. Here, we add string field call data to pass through the message.

```
using System;

namespace PaymentEngine.Internal.Commands.Payments
{
    9 references
    public class SubmitPayment
    {
        0 references
        public string data { get; set; }
    }
}
```

[73]

The Particular Service Platform

At this point, the code will not compile because the message only has one endpoint. We need to deploy the other endpoint with the **Deploy Component...** command, as shown here:

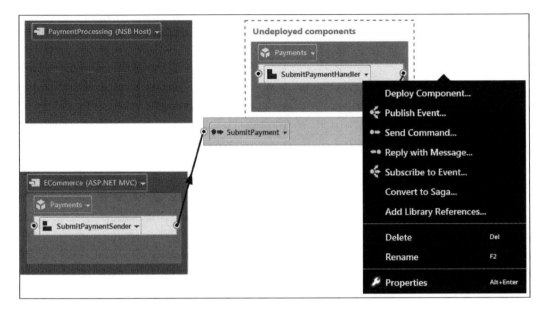

We can select the available endpoints. In this case, we have the ability to create new endpoints graphically, just as we did for the `PaymentProcessing` endpoint, as shown in the following screenshot:

Then, we will have two endpoints with a command message being sent from `ECommerce`, an MVC controller, to `PaymentProcessing`, an NSB Host. These endpoints will be command consoles or service applications depending on the deployment. In the following screenshot, we can see the two endpoints with the message in between:

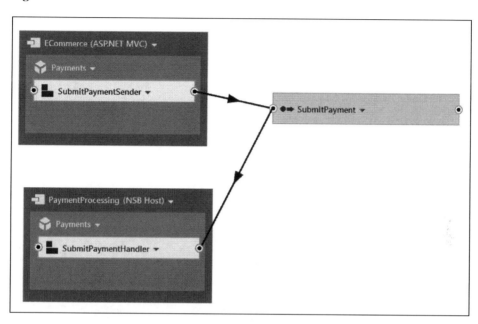

The `SubmitPaymentSender` function will send the message, and the `SubmitPaymentHandler` function will receive the message, as seen in the preceding diagram. These functions have already been created from ServiceMatrix and can be extended. Looking at `SubmitPaymentHandler`, we can extend the function to print the data field.

```
using System;
using NServiceBus;
using PaymentEngine.Internal.Commands.Payments;

namespace PaymentEngine.Payments
{
    1 reference
    public partial class SubmitPaymentHandler
    {
        2 references
        partial void HandleImplementation(SubmitPayment message)
        {
            // TODO: SubmitPaymentHandler: Add code to handle the SubmitPayment message.
            Console.WriteLine("Payments received " + message.GetType().Name);
            Console.WriteLine("Data " + message.data);
        }
    }
}
```

By running the project without adding any further code, we get a web-based interface to send the data in the message.

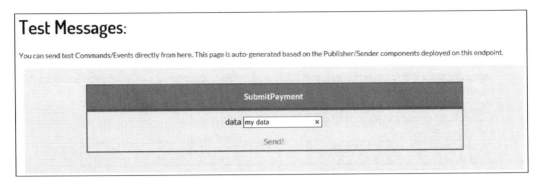

After sending the message, we receive the data that was sent in `PaymentProcessing`:

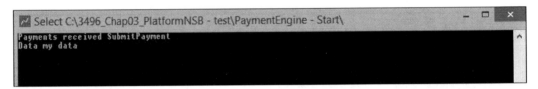

Let's add the plugins to ServiceMatrix. If we open up ServicePulse at `http://localhost:9090/#/dashboard`, we can see that the message appears at the two endpoints, but we need to install the plugin to monitor the endpoint. So we have basic endpoint messaging, but to provide more detail for the messages and endpoints, a plugin needs to be installed from ServiceControl. Here's some basic endpoint information in ServicePulse from this exercise:

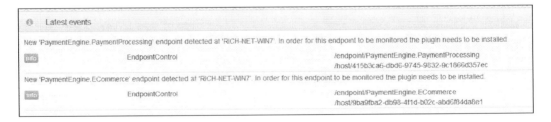

We will install the plugins for ServiceControl. After installing the plugins, if ServiceControl is not installed, you will receive an exception for ServiceControl. We can use the Package Manager Console to install the plugins as shown in the following screenshot:

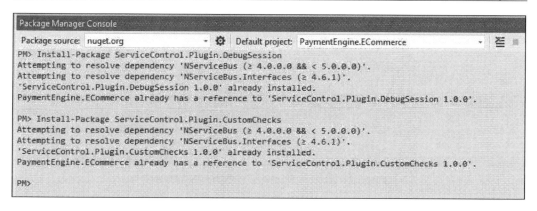

There are four ServiceControl plugins that can be installed, which we have mentioned so far:

- **ServiceControl plugin for CustomChecks**: The CustomChecks plugin allows the developer of an NSB endpoint to define a set of conditions that are checked on endpoint startup, or periodically.
- **ServiceControl plugin for DebugSessions**: DebugSessions is a dedicated plugin that enables integration between ServiceMatrix and ServiceInsight.
- **ServiceControl plugin for heartbeats**: The Heartbeat plugin sends heartbeat messages from the endpoint to the ServiceControl queue. These messages are sent every 10 seconds by default.
- **ServiceControl plugin for SagaAudits**: The Saga Audit plugin collects saga runtime activity information. This information enables the display of detailed saga data, behavior, and current status in the ServiceInsight saga view.

By installing the Heartbeat plugin into the `ECommerce` and `PaymentProcessing` projects, ServicePulse will now give heartbeat information on the uptime of these services, as shown here:

We can also run ServiceInsight to see the flow of the E-Commerce MVC sending the `SubmitPayment` to `PaymentProcessing`.

We can walk down the message and drill down for further information and insight into the performance and operation of the messages and endpoints.

Status	Message ID	Message Type	Time Sent	Critical Time	Processing Time	Delivery Time
♡	8ccb28be-4ee1-4ccc-9e5c-a38b01450da6	SubmitPayment	8/18/2014 7:43:28 PM	5s	171ms	5s

As you may notice, there are a few different times that are listed in the message properties, and they include the following:

- **Critical time**: The amount of time the message spends in transition from the sending to the processing endpoint.
- **Delivery time**: This is like the critical time, but includes waiting and processing time in the queue.
- **Processing time**: The amount of time it takes to actually process the message. This is done by the message processing handler method.

At this point, we should have a solution built in the `PaymentEngine - Start` directory that does basic request-response for a payment engine and publish-subscribe through ServiceMatrix.

The publish-subscribe messaging pattern is where the senders of messages, called publishers, will send messages without direct receivers. Instead, receivers of the messages and subscribers subscribe to the messages that they are interested in receiving.

Our wish is to end with the product having both publish-subscribe message flow and sagas. Also, these should end up with the solution in the `PaymentEngine - Saga` directory. NServiceBus, ServiceMatrix, ServiceControl, ServicePulse, and ServiceInsight were installed to walk through these scenarios.

Chapter 3

The **Publish Event...** option is used to create the message that will be published, as shown in the following screenshot:

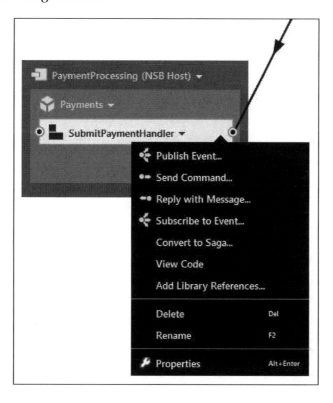

We will name the publisher event message `PaymentAccepted` from the `PaymentProcessing` host, as shown here:

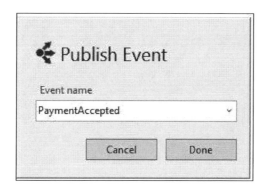

A code-convenient window will be created to review the code before deployment:

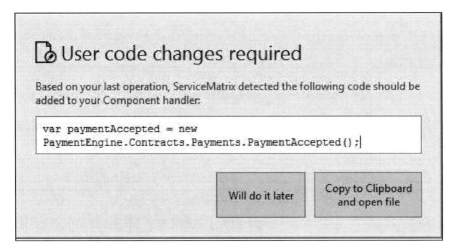

This is so that you can review the code before copying it into the message handler that you are publishing from. Here is the code:

```
public partial class SubmitPaymentHandler
{
    partial void HandleImplementation(SubmitPayment message)
    {
        // TODO: SubmitPaymentHandler: Add code to handle the
          SubmitPayment message.
        Console.WriteLine("Payments received " +
          message.GetType().Name);
        Console.WriteLine("Data " + message.data);
        var paymentAccepted = new
          PaymentEngine.Contracts.Payments.PaymentAccepted();
        Bus.Publish(paymentAccepted);

    }
}
```

Chapter 3

To add a subscriber to the publisher, simply use the **Add Subscriber...** command as shown in the following screenshot:

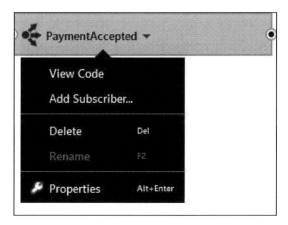

We can then add the subscriber to a new service. Let's call it `Paying`.

The Particular Service Platform

We will also have to deploy the `PaymentAcceptedHandler` component as an endpoint. In this scenario, we called it `Paying` as well. After these changes, we should see the following:

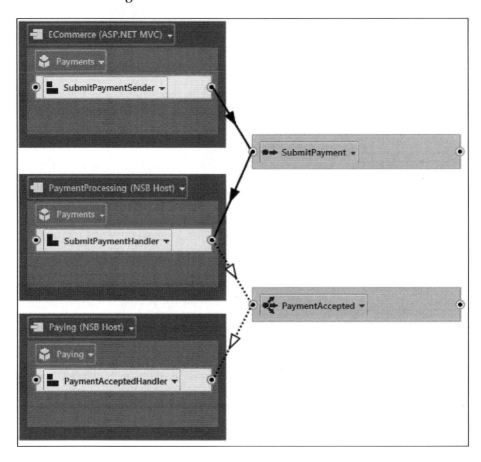

The **Properties** window of the solution will define the error and audit queues:

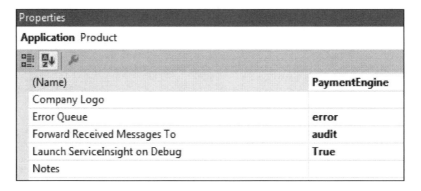

The **Properties** window will also show the various types of queues that can be used:

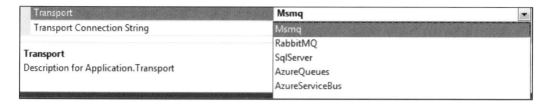

When running the solution and rerunning ServicePulse, we can see the additional `Paying` endpoint created, which hasn't had the plugins installed:

If we review the flow in ServiceInsight, we can see the new flows:

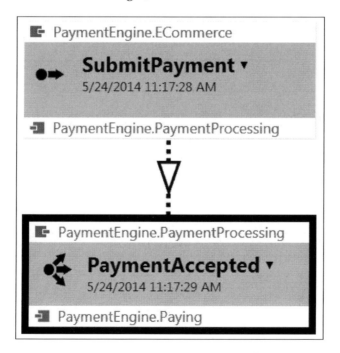

The Particular Service Platform

Sagas through ServiceMatrix

Not only can we develop endpoints for command messages and publish-subscribe messages, we can also develop sagas in ServiceMatrix. We will start by creating a new `PaymentNotification` command message:

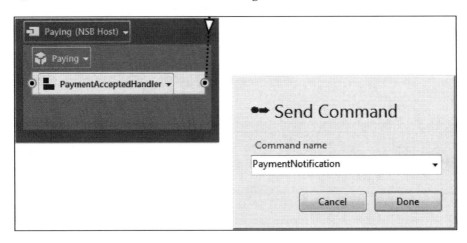

The following is the code for this piece:

```
public partial class PaymentAcceptedHandler
{
    partial void HandleImplementation(PaymentAccepted message)
    {
        // TODO: PaymentAcceptedHandler: Add code to handle
          the PaymentAccepted message.
        Console.WriteLine("Paying received " +
          message.GetType().Name);
        var paymentNotification = new PaymentEngine.
          Internal.Commands.Paying.PaymentNotification();
        Bus.Send(paymentNotification);
    }
}
```

Chapter 3

We will deploy the receiving endpoint to a new endpoint called `NotifyProcessing`.

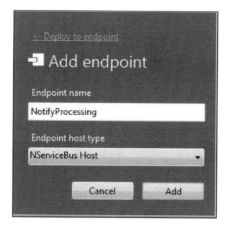

This is what we should have so far:

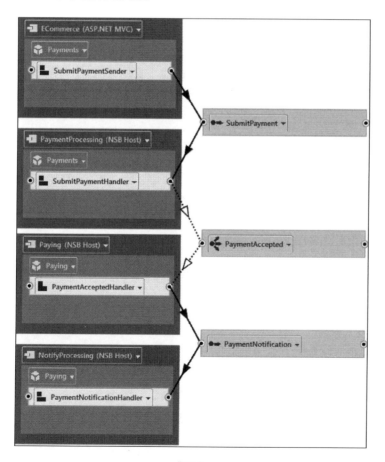

The Particular Service Platform

To start the saga process, we will use the **Reply with Message...** option, as follows:

This will allow us to convert the `PaymentAcceptedHandler` component into a saga.

After the saga is created, we can run the code from Visual Studio. Looking at ServiceInsight, we can see the updated flow diagram that contains all the endpoint components:

Chapter 3

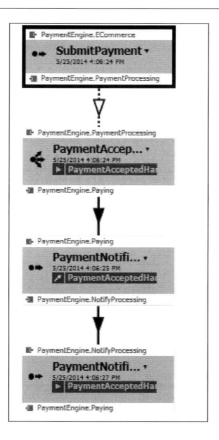

The ServiceInsight can display the flow of the various saga components. We can see the overall PaymentAcceptedHandler service to include which message will initialize the saga and which message will update the saga. Here we see the saga initiated with `PaymentAccepted`, saving `PaymentNotification` data, and updating the `PaymentNotificationResponse` message. This flow is displayed in the following screenshot:

Now, we have a solution that we saw while processing within ServiceInsight, which we could examine in the saga flow. We now have our solution in the `PaymentEngine - Saga` directory. NServiceBus, ServiceMatrix, ServiceControl, ServicePulse, and ServiceInsight were installed to walk through these scenarios. Because of the Particular NSB development tools, very little coding was needed to develop this application. However, we could examine the details of what was built in flow, message details, and have a graphical view in Visual Studio of the final solution.

Introducing CustomChecks for ServicePulse

With the `ServiceControl.Plugin.CustomChecks` plugin installed, we can perform several checks. In this section, we will be using the `PubSub--ReportFailure` solution—the `MyPublisher` project reports a failure check that will be reported in ServicePulse. This solution shows custom checks. In this section, we will also be using the `PubSub--ReportPass` solution—the `MyPublisher` project reports a pass check that will be reported in ServicePulse.

The following two functions can be used in the `CustomCheck` class using the `ServiceControl.Plugin.CustomChecks` plugin to generate a failure or pass condition to the ServicePulse dashboard:

- `ReportPass`: This will report that the custom check has passed.
- `ReportFailed`: This will report that a custom check has failed, passing in the string as the reason for the failure.

Here, we will create the code for a `CustomCheck` object that can be called when we submit a payment as an additional check. It is a simple constructor in a `MyCustomCheck` class that will pass information through its base class of `CustomCheck`. We called this class when we sent the `SubmitPayment` command from the `ECommerce` project using `MyCustomCheck myCheck = new MyCustomCheck();`. Here is the code:

```
using System;
using System.IO;
using ServiceControl.Plugin.CustomChecks;
using ServiceControl.Plugin.CustomChecks.Messages;
using ServiceControl.Plugin.CustomChecks.Internal;
namespace PaymentEngine.ECommerce
{
    public class MyCustomCheck : CustomCheck
```

```
    {
        public MyCustomCheck()
            : base("ECommerce SubmitPayment check", "ECommerce")
        {
            ReportPass();
        }
    }
}
```

So that when a submit payment is sent, we can get an additional message on ServicePulse.

ECommerce SubmitPayment check: Working as expected		
Info	CustomChecks	/customcheck/ECommerce SubmitPayment check /endpoint/PaymentEngine.ECommerce /host/9ba9fba2-db98-4f1d-b02c-abd6f84da8e1

We can use conditional statements to check for files that are present, other messages, and a number of conditions that can be reported as passing or failing while giving status to ServicePulse for operations to take action.

In the `CustomChecks` class, we can also set a timer to periodically check using the `PeriodicCheck` interface. This will set a timer to call back the class and send the condition to ServicePulse. It operates differently from ReportPass, as it is timer based in order to report the condition. When the new class references the `PeriodicCheck` interface, it requires a `PerformCheck()` function that will perform the custom check. In our function, we are using the NSB `CheckResult`

```
namespace PaymentEngine.PaymentProcessing
{
    class CheckHealth : PeriodicCheck
    {
        public CheckHealth()
            : base("PaymentProcessing Healthcheck",
              "PaymentProcessing", TimeSpan.FromMinutes(2))
        {
        }

        public override CheckResult PerformCheck()
        {
            // Fake a failure once in a while
            if (DateTime.Now.Second % 2 == 0)
            {
```

```
                    return CheckResult.Failed("PaymentProcessing fake
    failure");
            }
            return CheckResult.Pass;
        }
    }
}
```

This screenshot will demonstrate a failure condition in a custom check in ServicePulse:

This screenshot will demonstrate a pass condition in a custom check in ServicePulse:

There are many uses of custom checks in ServicePulse to give operations and the business the internal operations of the services, endpoints, and messages in NSB. Examples of custom checks include returning a failure or pass to determine if an e-mail server is running, an SFTP server is responding, a directory exists for saving files, a logging directory exists, and many more conditions. It would be good to report to ServicePulse so that operations are aware that a condition of the application will either pass or fail. We called this class when we passed messages to the `MyPublisher` queue using `MyCustomCheck myCheck = new MyCustomCheck();`.

We can then put in conditional statements to check for conditions and report a fail or pass. We can show how we pass a message to ServicePulse to report a pass. We can report a failure by replacing the report pass with a report failure, such as `ReportFailed("Testing")`. It will then log the failures in ServicePulse, as shown here:

ServicePulse provides a visual interface to show the history of the heartbeats, failures, and custom checks when it is running, and we can configure which available endpoints to check.

Summary

In this chapter, we looked at the various tools in the Particular Service Platform, which include ServiceMatrix, ServicePulse, and ServiceInsight. We had a very brief introduction of SeviceMatrix as we walked through building an E-Commerce MVC solution that works with request-reply messages using the send command. This was followed by publish-subscribe messages showing the ServicePulse and ServiceInsight results. ServiceInsight gives detailed information on each endpoint, message, and service, as well as a graphical flow to show the whole end-to-end flow. Of course, none of this could be done without ServiceControl, which is the service that is installed to collect the data to be sent to ServicePulse and ServiceInsight.

In the next chapter, we will discuss saga development. There will be a discussion as we look at saga development with web services using WCF, and using MVC as a frontend to read the various entries.

Saga Development

In this chapter, we will be focusing on the various useful constructions of sagas and message handlers. The purpose of sagas will be discussed when the need to extend and coordinate transactional integrity by using sagas is discussed. This chapter will then morph into a discussion of NServiceBus using integrated, pre-built WCF bridges. Some might consider it unusual to discuss WCF in a saga chapter, but sagas become an intermediary for coordinating WCF and NServiceBus workflows. We can decouple the workflow from the frontend for interaction to the backend processes through message handling. Sagas provide the means to persist the state information of the messages.

We will start with unit testing saga handlers and message handlers as we are constructing them, and how NServiceBus brings rules into testing them through Visual Studio. We will briefly discuss building our own tools and then move on to changing the transport mechanics from MSMQ into RabbitMQ. The goal is to know enough about developing sagas and message handlers so as to start building and testing our own sagas, and to have enough of an introduction to MVC, MSMQ, and EF at this point in order to start constructing and testing different business scenarios.

In this chapter, we will cover the following:

- A brief overview of MVC
- Sagas and web services
- Creating a WCF server
    - Messaging
    - Configuration tracing

Saga Development

- Creating a WCF client
    - Adding the service reference
    - Calling the reference
- Revisiting the design
- Adding the service reference
    - Calling the reference
- Adding NServiceBus to MVC
    - Message handler unit testing
    - Saga handler unit testing
- RabbitMQ for NSB
- ActiveMQ for NSB

A brief overview of ASP.NET MVC

Model-View-Controller (**MVC**) is the most common design pattern for implementing user interfaces. ASP.NET MVC is the Microsoft framework to implement the MVC software design pattern in ASP.NET. Developing by reusing known design patterns and frameworks that have been justified and tested by others brings a lot of reusability of known quantities into any application.

By breaking up the logic into controllers, which have the `session`, `request`, and `response` helper functions, while passing the models, which are the View Models which have the information which we want to present into the view, it moves most of the logic and exposure away from the browser where APIs can be exposed. We will also use Microsoft **Entity Framework** (**EF**) for many of the model objects.

Microsoft recommends using **Language Integrated Query** (**LINQ**) and EF to prevent traditional SQL injection attacks. The web page at `https://msdn.microsoft.com/en-in/library/bb308959.aspx` also discusses other security measures that can be done in EF. EF does not mitigate all injection attacks (such as EF injections) but, using a combination of EF and LINQ correctly, it will mitigate many common SQL injection attacks. The reasoning here is that the injection can now only occur through LINQ and the EF objects rather than any open SQL commands, thus narrowing the attack surface from a wide range of commands to an object, and through a collection that may only be accessed through a controller in MVC. There are many scripting tools running on the Internet, such as SQLNinja. Visit `http://sqlninja.sourceforge.net/` to find any insecure SQL command.

However, most tools are not built to signal EF attack vectors. Please visit http://www.slideshare.net/rhelton_1/sql-injection-amp-entity-frameworks for more information on this.

There are many more reasons to use an **object-relationship mapper (ORM)**, such as EF, as we can see in http://karwin.blogspot.com/2009/01/why-should-you-use-orm.html. Some of these reasons include the following:

- **Generating boilerplate code**: EF generates objects from the SQL Server databases and tables, thus creating boilerplate code that can be used to create, update, read, and delete the fields in the tables.

- **Supporting OO**: EF objects support common object-oriented programming design and methodology that is easy as PIE (polymorphism-encapsulation-inheritance), and to include reusability.

- **Speeding development**: Generating code from a database and using it in an application can be significantly faster than creating custom code from scratch. We can see the interaction of the MVC components in the following diagram:

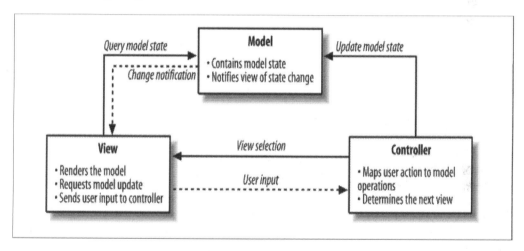

So, why is this discussion on MVC-EF in an NSB book? That's because most of the official NSB examples are using MVC. We will be looking at some examples that we have extended in MVC to include Microsoft's best practices of EF and WCF.

In order to view tables and queues related to tables, we use the Kendo grids. Packt publishing offers many books on Kendo, such as *Kendo UI Grid* (http://www.packtpub.com/kendo-ui-grid/book). Some of the Kendo examples are also extended from files available at http://www.codeproject.com/Articles/606682/Kendo-Grid-In-Action.

Sagas and web services

One of the many endpoints that NSB provide integration into is the **Windows Communication Foundation (WCF)** endpoint. Visit http://en.wikipedia.org/wiki/Windows_Communication_Foundation for more information.

WCF is part of the .NET Framework ecosystem which provides a runtime and a set of APIs for building connected, service-oriented applications. In the **Simple Object Access Protocol (SOAP)** binding, it makes use of **Web Services Description Languages (WSDL)**, which defines the interface between a web service and a web service client. Based on a WSDL, the XML is created on the WSDL specification so that the client and server can exchange information irrespective of their programming languages and platforms. It is sent between the server and the client as the protocol is normally HTTP or HTTPS. There are multiple binding types; we will discuss mostly SOAP binding in this book, but for more binding types, you can visit http://msdn.microsoft.com/en-us/library/ms731092%28v=vs.110%29.aspx. For example, if using a WCF client and service that are both built-in C#, we may consider using a NetTcpBinding class as it provides high performance between .NET WCF applications. Visit http://msdn.microsoft.com/en-us/library/ms731092%28v=vs.110%29.aspx for more information.

For simplicity's sake, we are only going to work with a SOAP web service in WCF in this chapter.

WCF is a highly extensible framework and allows easy integration into NServiceBus. More information on NSB WCF sample integration can be found at https://github.com/Particular/NServiceBus/tree/develop/IntegrationTests/WcfIntegration and http://docs.particular.net/NServiceBus/how-do-i-expose-an-nservicebus-endpoint-as-a-web-wcf-service.

While Microsoft's WCF is considered a framework for implementing pieces of **service-oriented architecture (SOA)** guidelines, it is not an ESB. This WCF does not contain all the features and design patterns (like sagas), persistence, or other out-of-the-box features to implement an end-to-end solution for SOA guidelines. On the other hand, using NServiceBus as an ESB with the web services of WCF brings a lot to the table that WCF and other web service frameworks do not offer.

The source code

The directory for the code is under the Payment_WCFService directory. The WCFService is used to test the client by being the WCF service.

Chapter 4

The solution is then run in Visual Studio 2012 in Windows Server 2012 with MSMQ, DTC, RavenDB, NServiceBus Version 4.0 references, and SQL Server 2012 Express LocalDB installed.

The WCFService must first be running for the client, Payment_WCFService, to send it messages.

Creating a WCF server

We will start by creating a WCFServer project in Visual Studio, as the server needs to be running before it can communicate with the client. By adding the reference NServiceBus.Host from NuGet or Package Manager Console, several NSB default settings will be created in the App.config file for the project, an EndpointConfig.cs file with AsA_Server will be created, and the project will be set to run with an NServiceBus.Host.exe executable. We can see the creation of the files by the reference in the following screenshot:

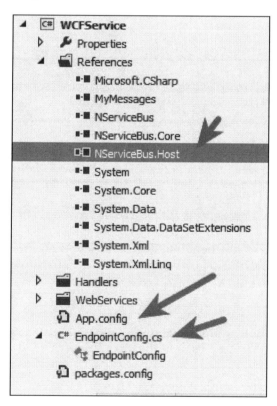

Saga Development

When we add an `NServiceBus.Host` reference, through either NuGet or Package Manager, into the project, many other items will be added into the project as well. These other items are as follows:

- An `EndpointConfig.cs` file will be created in the project with the default settings to add endpoints.
- The project will be set to run as a DLL, being executed by an `NServiceBus.Host.exe` executable when run from the debugger.
- Several default settings for creating a generic `AsA_Server` endpoint will be added to the `App.config` file.

We can see from the following screenshot that when we add the `NServiceBus.Host` reference, this sets the project to run as a DLL using the `NServiceBus.Host.exe` executable:

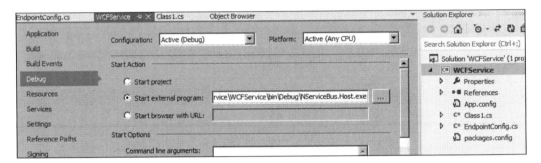

We will use NSB WCF Integration to save a lot of work on our part. In order to use this integration, we will need a few pieces related to both NSB and WCF. These are as follows:

- **A web service**: This is used with the `NServiceBus.WcfService<TRequest, TResponse>` class where we define a request and response on the web service. We will show here that the service called `PayService` will receive the `PaymentMessage` message as the request and respond with `ErrorCodes`. This is shown in the following line of code:

  `public class PayService : WcfService<PaymentMessage, ErrorCodes>`

- **A message handler**: This is required to handle the message from the web service and to process it in the integration. We will call this handler `PayHandler`.
- **The request and response message structures**: This is required for the `PaymentMessages` project.
- **The configuration for NSB with WCF**: This is required in the **App.config** file.

Using NSB with WCF integration is similar but different from using straight WCF. In WCF, there are four basic steps that can be extended. Visit `http://www.c-sharpcorner.com/UploadFile/dhananjaycoder/four-steps-to-create-first-wcf-service-for-beginners/` for more information.

These steps to create a WCF server are as follows:

1. Create a service contract. This defines the available functions between the WCF client and the WCF server through interfaces.
2. Expose endpoints with metadata—through either `App.config` or `Web.config`. We expose the endpoints through the configuration to include how the exchange of data will occur.
3. Implement the service. That is, we add functionality and data objects to the interfaces.
4. Consume the service. Or, in other words, expose the service to be imported into a WCF client.

In using NSB with NHibernate, NSB takes care of the mapping of NHibernate with the messages, endpoints, and services. In WCF integration, NSB also takes care of many of the service contracts with the use of the message handler and the message format.

We can observe these features pieces in the following screenshot and we will discuss them further:

We will add the messages next.

Adding messages

We will now create a `PaymentMessages` project inside the solution, as shown in the following screenshot:

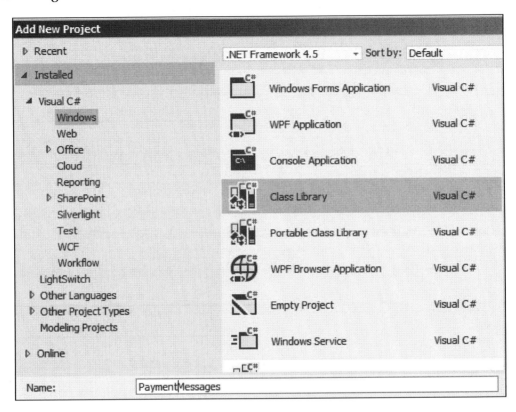

We will be using the `PaymentMessage` request message, which the request needs to be NServiceBus IMessage Interface having two classes, a `Guid` method for a unique ID per message, and a `PaymentReq` class:

```
public class PaymentMessage : IMessage
{
    public Guid EventId { get; set; }
    public PaymentReq paymentReq { get; set; }
}
```

The `PaymentReq` class will have many fields which are necessary for a normal payment to a bank, including items like bank routing number, bank account number and many more items to identify a payment (as shown in the following diagram):

The `ErrrorCodes` are just an `Enum` class that either returns `None` (for no error) or `Fail`, as shown in the following screenshot:

Now that we have a message, and we also have a WCF service (called `PayService`), let's start a message handler to handle the message by using the message to receive it from the client.

Just to give an overview, `PayService` will get a message from the web service client and, in turn, place the message in the MSMQ by default (that is, a queue called `wcfservice`) for the message handler to process it, and then respond back to the web service.

Adding the message handler

We will now create a message handler to handle the message (`PaymentMessage`). Here, we are just printing out the `EventId` instance to the console window. Further, we may add subscription processing for handling the payment as a bank may handle it. The service will return an `ErrorCode` with the value of `None`. We are conducting very simplistic tests at the moment. This is the WCF service project under `BasicWCF1`:

```
namespace WCFServer.Handlers
{
    public class PayHandlers : IHandleMessages<PaymentMessage>
    {
        private readonly IBus bus;
        public PayHandlers(IBus bus)
        {
            this.bus = bus;
        }
        public void Handle(PayMessage message)
        {Console.WriteLine("=======================================
     ===============================");
            Console.WriteLine(message.EventId);
            Console.WriteLine("==================================
        ==================================");
            bus.Return((int)ErrorCodes.None);
        }
    }
}
```

We will then create the WCF client for the web service call. The client will simply define the messages to send to the web service for handling the request and response. The `PayMessage` instance will be the request going from the client to the service. The reply from the service will be the `ErrorCodes`. We will perform these steps after we explore more about the configuration and tracing of the WCF service.

Adding the configuration

The task of configuring the `App.config` file for the web service still remains. This step is similar to *exposing the endpoint with metadata* that we mentioned earlier.

The `App.config` file will define several characteristics of the web service, such as the listening port and URL, the security of the service, and the binding type for the service. The binding type defines the communication mechanism of the endpoint, be it basic HTTP, MSMQ, or any other. A list, as well as more information on WCF binding, can be found at `http://msdn.microsoft.com/en-us/library/ms730879(v=vs.110).aspx`.

We could configure the `App.config` file by manually editing it, or by using the WCF Service Configuration Editor that comes as part of Visual Studio.

More information on the Configuration Editor can be found here:

`http://msdn.microsoft.com/en-us/library/ms732009(v=vs.110).aspx`

We will now open the `App.config` file through the Configuration Editor to do the following:

- Establish a server URL, including port number, to be available for the client WCF as the WCF endpoint
- Establish the binding parameters that will expose the WCF service endpoints in a variety of different ways
- Set up tracing and logging to review the transmissions and services as they happen

Saga Development

We start by opening the project's App.config file and setting up the URL and ports that the web service will be listening with to start configuring the endpoint—as shown in the following screenshot:

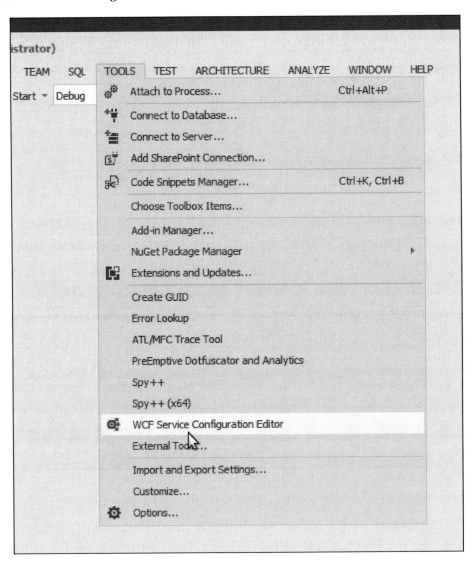

Chapter 4

Next we set the binding types. We will be using the WCF server endpoint to be set at a particular port, as shown in the following example:

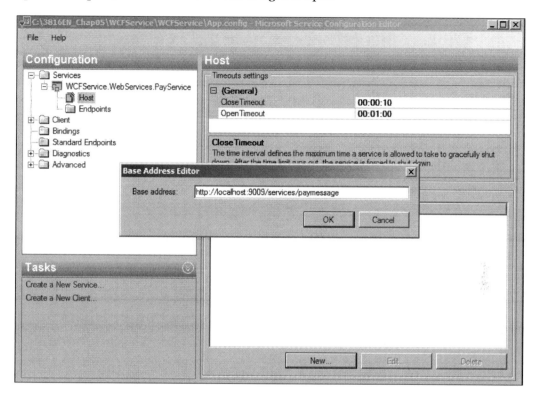

We used mex binding here, which is metadata exchange binding. This is useful if data needs to change over time and the update of client information through discovery. In other words, we are adding metadata to the endpoint to expose the metadata of the service so that the WCF client can easily create a proxy.

Please visit `http://msdn.microsoft.com/en-us/library/ms731734(v=vs.110).aspx` to review the WCF configuration schema. We can edit the binding through the service endpoint as shown here:

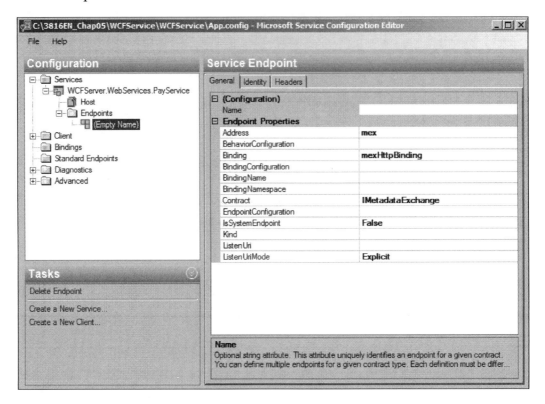

Adding tracing

There may be some debate as to whether this section is required or not, as this section is specific to WCF but not related to sagas or NSB. However, if your WCF communication is not working between a WCF client and a WCF server, the issue could be related to the message structure, the network, the WSDL, or any other factor. And in order to trace the issues in WCF, the tracing feature needs to be enabled and set up. This also allows us to see the reaction of the communication. However, if your WCF is functional and no troubleshooting is required, feel free to skip this section.

There are service tracing utilities within WCF to view graphs, messages, network calls, exceptions, and activities of WCF web services and web clients. Please refer to `http://msdn.microsoft.com/en-us/library/ms732023(v=vs.110).aspx` for more information on Microsoft Service Trace Viewer.

We can capture the messages between a client and a server by setting the diagnostic sections (given as `<diagnostics/>`) in the `App.config` file to capture these messages, and also by setting the listeners and sources, which will be associated with the libraries, to log events in the `App.config` file, as shown here:

```xml
<AuditConfig QueueName="audit" />
<system.serviceModel>
    <diagnostics>
        <messageLogging logEntireMessage="true" logMalformedMessages="false"
            logMessagesAtServiceLevel="true" logMessagesAtTransportLevel="true"
            maxMessagesToLog="30000" maxSizeOfMessageToLog="30000" />
        <endToEndTracing activityTracing="true" messageFlowTracing="true" />
    </diagnostics>
    <behaviors>
      <serviceBehaviors>
        <behavior name="Default">
          <serviceMetadata httpGetEnabled="true" />
          <serviceDebug includeExceptionDetailInFaults="true" />
        </behavior>
      </serviceBehaviors>
    </behaviors>
    <services>
      <service name="WCFServer.WebServices.PayService" behaviorConfiguration="Default">
        <endpoint address="mex" binding="mexHttpBinding" contract="IMetadataExchange" />
        <host>
          <baseAddresses>
            <add baseAddress="http://localhost:9009/services/paymessage" />
          </baseAddresses>
        </host>
      </service>
    </services>
  </system.serviceModel>
</configuration>
```

Saga Development

The following screenshot represents how we can configure message logging and what kind of trace listener we are going to use to write the logs:

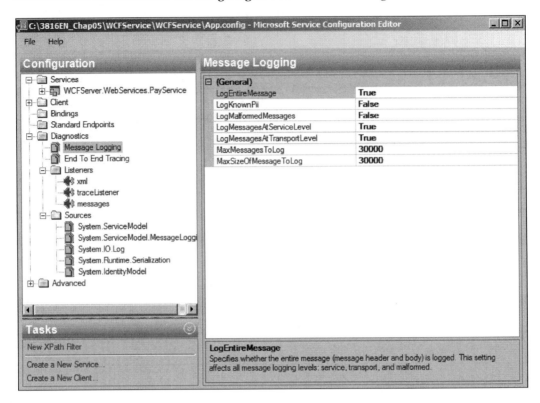

We can view the messages being sent between the client and the server by opening the `svclog` file in Service Trace Viewer. As shown in the following screenshot, we can see the SOAP message, including the header and the body, which is what the WCF server was receiving.

Chapter 4

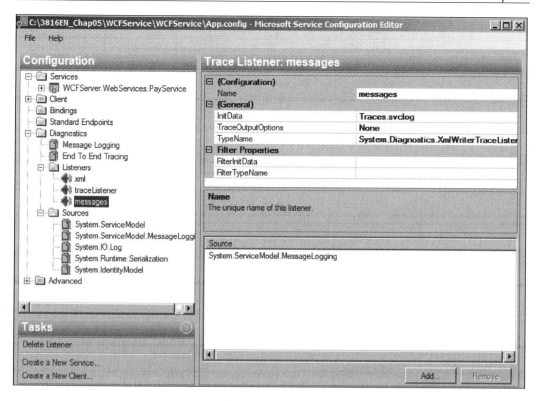

Not only can we see the SOAP information, but the HTTP information as well. By adding a `System.Net` listener, we can trace the network socket calls for opening and closing sockets as they happen, as well as the sessions in between. For additional information on how to configure network tracing for WCF, please refer to the following links:

- http://msdn.microsoft.com/en-us/library/ty48b824%28v=vs.110%29.aspx
- http://msdn.microsoft.com/en-us/library/ms733025%28v=vs.110%29.aspx

So, why discuss web service tracing? In many web services, such as those connecting to banks to process credit cards, we may only see one side of the WCF service. Sometimes, the other side may end up being a Java web service or other types of services that a third party develops; it may even be using CICS legacy code. Consequently, we may have very little control over some of the web services that we will be integrating into. And sometimes, there is very little documentation as well. So, tracing becomes a necessity for many web services to debug and log functionality as messages go in between systems.

Saga Development

Viewing the web service

Running the web service in Visual Studio 2012, we can view the web service through the browser at `http://localhost:9009/services/paymessage`. It will provide some simple instructions for the WCF client, as we can see in the following screenshot:

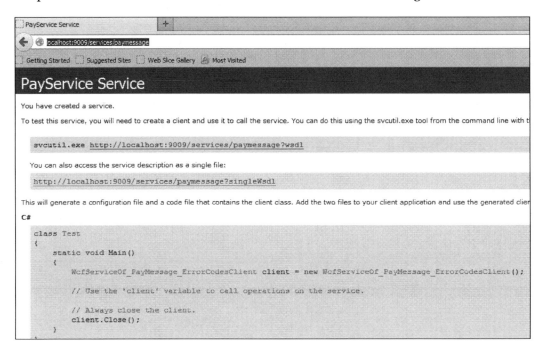

And we can also see the WSDL by accessing the page from `http://localhost:9009/services/paymessage?singleWsdl`:

Considerations when deploying

Please note that for most production systems, it is always recommended to install digital certificates when using web services to encrypt the pipeline and verify the identity of the client and the server.

For a real-life production system, all the communication between the WCF client and WCF server must be secure using SSL, and you should look into your web services using WS-Security-based applications.

Creating a WCF client

In this section, we will create an `MVCApp` solution that will be under the `BasicPayClient` directory. It will contain several projects as follows:

- `MVCApp`: This contains MVCs, Kendo grids, and the WCF client for a browser user interface to send messages to the WCF service. It will read five XML files to load as messages.

- `MyMessages`: This contains `IMessages` of NServiceBus for the building of messages.

- `WriteXMLFiles`: This is a utility to write five XML files to a `C:\temp\` directory for the `MVCApp` project to load. This application saves messages in the form of XML files, which in turn are loaded through `MVCApp` to be sent from the WCF client to the WCF service. These are for testing purposes only, but using files in the form of messages makes it quick to change the data for various tests in the communications and endpoints. The messages are read from a `C:\temp\` directory as five XML files saved in a format that works with the WCF messaging service. The files can be created in the `C:\temp\` directory by running the `C:\temp WriteXMLFiles` project. These files are simply test messages and are saved to a disk so that they can be modified and tested easily.

Next, we need to design the client. We will have an MVC ASP.NET web interface that reads the XML files and displays them. We can select an individual message and send it to the WCF service as a request to get a response.

As a further exercise in this chapter, we will add a saga and a message in-between the MVC application and the WCF client to show how the workflow will assist us.

We already have instructions in the display of the WCF service, as given here:

To test this service, you will need to create a client and use it to call the service. You can do this using the svcutil.exe tool from the command line with the following syntax:

svcutil.exe http://localhost:9009/services/paymessage?wsdl

You can also access the service description as a single file:

http://localhost:9009/services/paymessage?singlewsdl

> In order to verify if the service is up and running, just browse to your service at the following link:
> `http://localhost:9009/services/paymessage`

Adding the service reference

First, we will run the `WPayment` WCF service from Visual Studio. We can check to see whether it is running by just seeing whether we can view the web service in a browser. This is a self-hosted solution since we can use NSB to deploy the service as it is contained in an NSB package. However, WCF can be IIS-hosted as well. To see a comparison of some of the Windows hosted solutions, visit `http://msdn.microsoft.com/en-us/library/ms730158.aspx`.

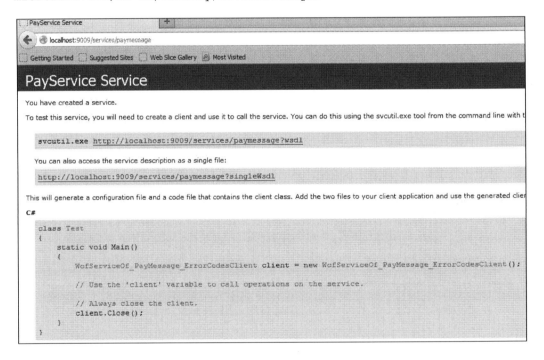

Chapter 4

We will now add a service reference from the available WSDL to the web service client. Visit `http://msdn.microsoft.com/en-us/library/bb628652.aspx` for more information. The following screenshot shows how the **Add Service Reference** window appears:

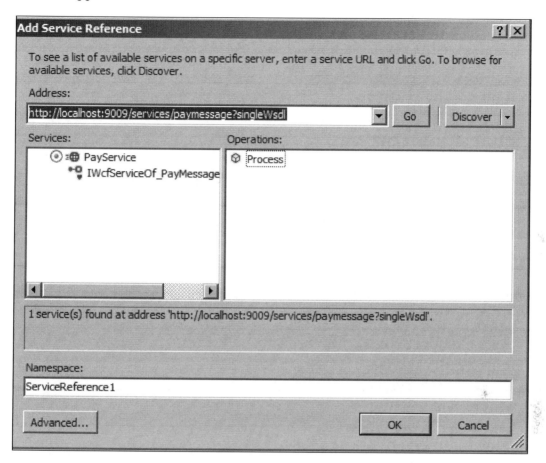

Saga Development

We will use the advanced settings to reuse the `MyMessages` packages of messages that we are using in this chapter's projects. The following image shows how your screen would look:

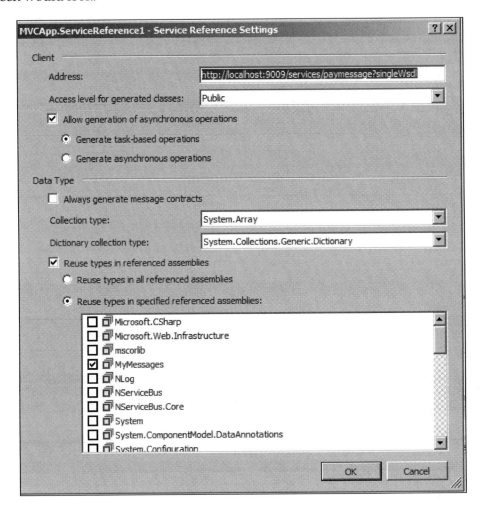

Calling the service reference

We will create an `index.html` file on the Kendo grid which will be a link on **View** which, when clicked, will load the XML files. When a particular XML `PayMessage` instance is selected to be sent to the WCF service, it will call the service reference, which was imported as `ServiceReference1`, to create a client and pass the selected `PayMessage` into it to be processed and sent to the WCF service. We will create this code in the MVC controller function, `SendWCF`, and pass it the ID of the message that we are sending to the WCF service. The code is shown in the following screenshot:

```csharp
public ActionResult SendWCF(int id)
{
    var user = new XMLLoads().GetPayments().Where(p => p.Id == id).FirstOrDefault();

    var message = new XMLLoads().GetMessages().Where(p => p.EventId == user.EventId).FirstOrDefault();

    ServiceReference1.WcfServiceOf_PayMessage_ErrorCodesClient client1 =
        new ServiceReference1.WcfServiceOf_PayMessage_ErrorCodesClient();

    ErrorCodes returnCode = client1.Process(message);

    user.errorCode = returnCode;

    return View(user);
}
```

This will be the Kendo grid in the browser that offers a selection to be sent to the WCF service. Upon browsing the page, the grid will populate with the following data and an option to send the message to the WCF service, as shown in the following table:

Event Id	Action
8b265223-dc9e-4789-a6df-69d19f644ad7	Call_Payment_Service
3721ba5d-4733-4d98-a5e2-8e8afa3e61f4	Call_Payment_Service
1ac188ec-4b2e-436c-b989-db88c65db1fa	Call_Payment_Service
9bf180fa-f8f4-4b2b-8fac-cca73a4e2cab	Call_Payment_Service
ee2c56f7-6d42-4314-bce5-4825ed294437	Call_Payment_Service

The Kendo grid scripting inside the `SendWCFPay.cshtml` file will look like the following:

```html
<h2>Send Payment Message to WCF</h2>
    <script type="text/javascript">
        $(document).ready(function () {
            var modelData = @Html.Raw(Json.Encode(Model))
            $("#grid").kendoGrid({
                pageable: true
                , sortable: true
                , silectable: true
```

Saga Development

```
            , selectable: true
            , columns: [
            { field: "Id", title: "Id" }
          , { field: "EventId", title: "Event Id" }
          , { title: "Action", template: '<a href=
              "/user/sendwcf/#=Id#">Call_Payment_Service</a>' }
            ]
            , dataSource: { pageSize: 10, data:modelData }
        });
    });
</script>
```

If the `PayMessage` was processed successfully, it will have no errors when we view the details of that request, as shown here:

Revisiting the design

In the code, as an example, we have built a `Payment` WCF Service solution as a WCF Integration solution into NSB. We have also built an MVC frontend to the WCF client solution that we can use for testing the Payment WCF Service.

The frontend is a basic MVC application with some basic WCF client interfaces for a `PayMessage`. `PayMessage` has a GUID, an address, and basic information for payment.

Chapter 4

For simplicity, the MVC controller is just reading XML files (that were created in a folder in `C:\temp\`) to be displayed in the frontend and selected to be sent to the Payment WCF service. Using XML files in this method is not recommended for production, as normally these files would be filled in by a frontend payment interface to the customer for the customer information. We chose this method to solidify the interfaces and test them. And designing customer interfaces to populate the messages was not part of this effort. Using XML files in this type of testing can be advantageous, as we could extend this example to send hundreds of messages, both in order to check performance as well as to populate individual messages through a customer interface without relying on the frontend.

What we basically have is a very common design, where we have an MVC frontend calling a WCF client to interact with a WCF server. This design is also common to see at some ordering sites. For instance, at, let's say a pizza site, is a pop-up saying **order being sent, do not refresh**. Because there is no decoupling to the backend workflow in this example as well, if this were to become a solution without decoupling through NSB, we may likely have to add payment being sent, do not refresh. Many things can go wrong in the event of during the need for a page refresh, just because a customer may not be refreshing their page, it does not mean that something else isn't doing a page refresh.

Our current architecture is as follows:

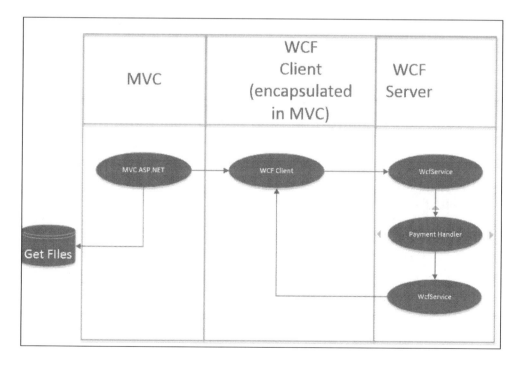

Saga Development

Let's build some saga processing into the preceding solution to decouple the MVC frontend from the WCF client. NSB will act as a mediator for the interaction between these two components pieces.

Why a saga? It routes messages, performs timeouts, and persists a state.

The preceding screenshot represents the new workflow using the NServiceBus saga. It seems a bit overwhelming, but NServiceBus handles all the complexity behind the scenes. For the controller, it just fires and forgets. If it needs to know the state of the message, it just executes an EF query on the DB. But it still sends the message off, and allows NSB to handle it.

The source code

The directory for the code is under the `SagaPaymentClient` directory.

The MVCApp – WCF is used to send WCF messages as a client using a saga.

The solution was built using Visual Studio 2012 in Windows Server 2012, and also tested with VS 2012 running in Windows 8.1, with MSMQ, DTC, RavenDB, NServiceBus Version 4.0 references, and SQL Server 2012 Express LocalDB installed.

Here's what the new project will look like with the saga:

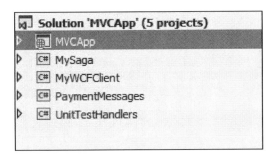

The difference now is that the MVC application will interface with the saga, which is `MySaga`, and the saga will communicate with the WCF client and the MVC application. This will decouple the frontend from the WCF client, as well as have the MVC receive feedback of the status directly from the saga message.

The interaction will look like the following:

Wow! It looks like a lot, but most of the diagram is similar to the previous MVC application, and we are familiar with most of the orchestration represented in the preceding diagram. However, all the new pieces of orchestration are handled by the NSB saga. The WCF client and server remain the same, but the WCF client is encapsulated in a message handler that receives and responds to the saga. Before refactoring with NSB, the WCF client updates the MVC application with the status of the WCF by updating a database about the fact that the message was processed successfully from the WCF server. Now, after NSB, the MVC application has a message handler. It will still update the database, but now the MVC application can be event driven from the saga as well, whenever the user needs to be notified immediately. Has the saga refactoring process increased the quality of software? If a message is interrupted, the power is shut off, the user refreshes their page, and a thousand other things could happen. However, the payment will not be lost. Also, a page refresh will not affect the payment either.

As part of this refactoring into NSB, the saga handles all the complexity around the service calls, and we have a nice separation of concerns between the controller and the orchestration.

Adding NServiceBus to MVC

We will next be extending our `BasicPaymentClient` folder and projects into a `SagaPaymentClient` folder and projects with the addition of the bus. The differences will be as follows:

- We will refactor the WCF client out of the MVC controller and move it into a new message handler.
- We will create a database that keeps track of the state of the message. The saga data will be saved in the `nservicebus` table. Ensure that it is created when running the saga code.
- We will create a new message handler and put the WCF client code in it. This code from the WCF client to the WCF server will be kept separate from the frontend code as far as possible, to keep the `PayMessage` class completely separate from the frontend.
- We will create a new message handler in the MVC that will update the DB with the message state as it receives the new state from the endpoints. We will look at the code for the message and saga handlers as we test them in the next section.
- `NServiceBus.Testing` offers testing by sending messages through message handlers and sagas. This includes anything that a message handler and saga can do, including header manipulation and dependency injection. Refer to `http://docs.particular.net/NServiceBus/unit-testing` for some basic examples. For the source code of `NServiceBus.Testing`, go to `https://github.com/Particular/NServiceBus/tree/develop/src/`.

The very basis of starting unit testing is to create a unit testing project in Visual Studio by adding a new unit testing project to an existing solution. Visit `http://msdn.microsoft.com/en-us/library/hh598957.aspx` and hyperlink it. The **Add New Project** window is as shown here:

Chapter 4

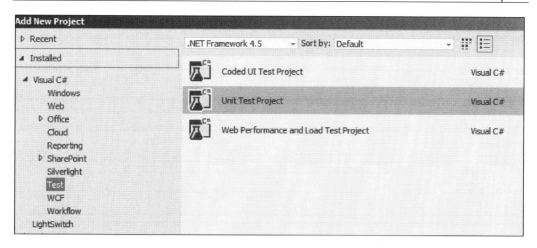

We will add the `NserviceBus.Testing` project from NuGet (http://www.nuget.org/packages/NServiceBus.Testing/). Your screen should look similar to what is shown in the following screenshot:

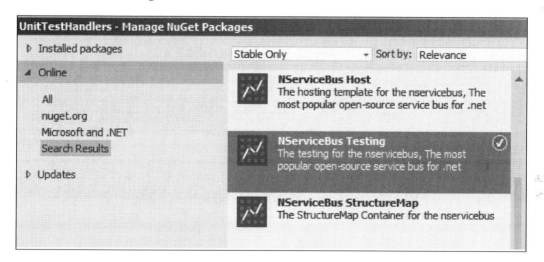

Saga Development

In the `NServiceBus.Testing` projects, all the tests are initialized with the `Test.Initialize()` method. A test will originate with the `Test.Handler<HandlerName>()` or `Test.Saga<SagaName>()` methods. This is shown here:

```
using NServiceBus;
using NServiceBus.Testing;

namespace UnitTestHandlers
{
    [TestClass]
    public class UnitTestHandler
    {
        /***
        *
        * Test the message handler for MYWCFClient
        * This will call the WCF Service for a completion
        *
        * ****/
        [TestMethod]
        public void Run()
        {
            Test.Initialize();

            Test.Handler<MyHandler>()
                .ExpectReply<ResponseMessage>(m => m.String == "hello")
                .OnMessage<RequestMessage>(m => m.String = "hello");
        }
```

When a test is built, we can run it or debug it. The test indicators will tell us if anything failed or succeeded. As part of following **test-driven development** (TDD), we must follow the AAA rule. These rules incorporate the Arrange-Act-Assert (AAA) pattern to verify whether a test fails or passes. Visit `http://c2.com/cgi/wiki?ArrangeActAssert` for more information.

We can also put in rules and assertions where, if the correct response does not happen, it will fail the test. This is a great feature of Visual Studio, and there are many samples at `http://msdn.microsoft.com/en-us/library/ms243176.aspx` and `http://www.visualstudio.com/en-us/get-started/create-and-run-unit-tests-vs.aspx`. There are extensions available as well at `http://www.codeproject.com/Articles/22358/Visual-Studio-Unit-Testing-Extensions`.

Message handler unit testing

The message handler code will be in the unit test itself. From our project, `UnitTestHandlers`, in which we have various unit tests, we will walk through `EventMessageHandler`. `EventMessageHandler` receives a `SendCommand` object from the MVCApp, via the saga, as shown here:

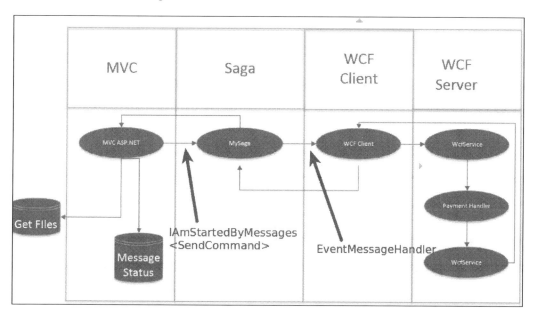

We will proceed with creating a `UnitTestHandler2.cs` file, and then add the header information and `[TestMethod]`. This will be under `SagaPaymentClient` in the `UnitTestHandlers` project.

Saga Development

After the base of the file is created, we will create a normal message, `SendCommand`, with a GUID and state where the message should be at before reaching the message handler, called `command`. The code is as follows:

```csharp
namespace UnitTestHandlers
{
    [TestClass]
    public class UnitTestHandler2
    {
        /***
        *
        * Test the message handler for MYWCFClient
        * This will call the WCF Service for a completion
        *
        * ****/
        [TestMethod]
        public void Run()
        {
            Test.Initialize();
            /*****
            *
            * Create a Command message
            * used to look up an XML Message file
            * on Disk, send to WCF Server
            *
            * *****/
            SendCommand command = new SendCommand();
            command.RequestId = new Guid("8b265223-dc9e-4789-a6df-69d19f644ad7");
            command.state = MyMessages.MessageParts.StateCodes.SentMyWCFClient;

            // The Test code
            Test.Handler<EventMessageHandler>()
                .ExpectReply<ResponseCommand>(m => m.state == MyMessages.MessageParts.StateCodes.CompleteMyWCFClient)
                .OnMessage<SendCommand>(command);
        }
    }
```

We see that the command message is passed into the `.OnMessage<SendMessage>(command)` method and a `ResponseMessage` object in the `Reply` method, with the state being set to `CompleteMyWCFClient`. When calling the unit test in **Debug**, we can even pass this message in the handler and see how it behaves, as shown here:

Chapter 4

```
public class EventMessageHandler : IHandleMessages<SendCommand>
{

    public IBus Bus { get; set; }

    public void Handle(SendCommand message)
    {

        ServiceReference1.WcfServiceOf_PayMessage_ErrorCodesClient client1 =
            new ServiceReference1.WcfServiceOf_PayMessage_ErrorCodesClient();

        // Create the response message
        ResponseCommand command = new ResponseCommand();
        command.RequestId = message.RequestId;
        /****
         * Get the XML messages from the temp direcotry.
         * Find a match from the GUID
         * ****/
        List<PaymentMessage> list = EventMessageHandler.GetMessages();
        PaymentMessage payMessage = null;
        foreach (var temp_message in list)
        {
```

Name	Value	Type
⊟ command	{PaymentMessages.ResponseCon	PaymentMessages.ResponseCommand
⊞ Request	{8b265223-dc9e-4789-a6df-69d19	System.Guid
state	initial	PaymentMessages.MessageParts.StateCodes

This allows us to design and debug the handler functionality in the unit test code through TDD. There are many rules that can be used when testing the handler or saga. For instance, ExpectNotReply is used to expect that the handler does not reply with a specific message.

To get information on what is available in NServiceBus.Testing, we can execute the following steps:

1. Try to enter something and hover over IntelliSense.

Saga Development

2. Read the documentation at `http://www.nudoq.org/#!/Packages/NServiceBus.Testing/Handler%28T%29`. The following screenshot shows the documentation:

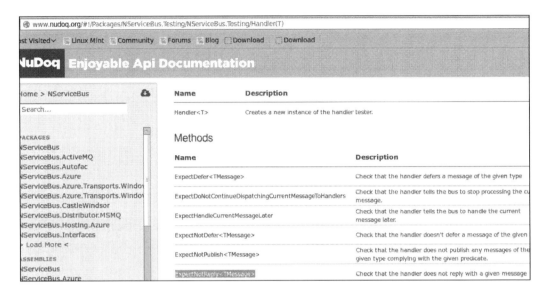

3. Read the code in GitHub at `https://github.com/Particular/NServiceBus/blob/develop/src/NServiceBus.Testing/Handler.cs`:

So, there are many possibilities for testing to test the code. For the message handler, it will get the command with the GUID and state, read the XML files to get a matching message, and send it to the WCF service, which will respond back to the saga. The saga keeps track of the message routing and states, and will respond to the MVCApp. The MVCApp will consequently update its state in the table. There could normally be multiple Views that could read the state – maybe an admin utility to check on the state of the messages, the **customer service rep** (**CSR**) talking to the customer, telling them if the payment has been processed, or a confirmation form or e-mail to the customer telling them that the payment succeeded, or many other scenarios.

Chapter 4

Besides a couple of functions to read the XML file for the message, which is just used for testing, there could be a number of scenarios added, but the majority of the code to do this is simply the following. Simple enough!

```
/****
    * The message handler
    * Matches an XML message GUID from a file and the command
      sent
    * to it from MVC via the Saga
    * If found, sends it to the WCF  Server and responds
    * with the state of what happened.
    * The WCF Service must be running to complete.
    *
    * ****/
public class EventMessageHandler :
  IHandleMessages<SendCommand>
{
    public IBus Bus { get; set; }
    public void Handle(SendCommand message)
    {   ServiceReference1.WcfServiceOf_PayMessage_
        ErrorCodesClient client1 =
             new ServiceReference1.WcfServiceOf_
             PayMessage_ErrorCodesClient();

        // Create the response message
        ResponseCommand command = new ResponseCommand();
        command.RequestId = message.RequestId;
        /****
          * Get the XML messages from the temp directory.
          * Find a match from the GUID
          * ****/
        List<PayMessage> list =
           EventMessageHandler.GetMessages();
        PayMessage payMessage = null;
        foreach (var temp_message in list)
        {
            if (message.RequestId == temp_message.EventId)
            {
                payMessage = temp_message;
            }
        }

            ErrorCodes returnCode =
              client1.Process(payMessage);
```

[127]

```
                if (returnCode == ErrorCodes.None)
                {
                    command.state =
                        StateCodes.CompleteMyWCFClient;
                }
                else
                {
                    command.state =
                        StateCodes.MyWCFClientFail;
                }

                Bus.Reply(command);
                Console.WriteLine("Success");
            }
```

What happens if the XML file does not exist? The following is the code used then:

```
// if no XML, just fail
            if (payMessage == null)
            {
                command.state = StateCodes.MyWCFClientFailXML;
                Bus.Reply(command);
                Console.WriteLine("No XML Found");
            }
            else
            {
               ... normal path   }
```

After testing this code, we could use the tested code to create a class into a new project, barring the unit testing, and start using it as a message handler. It saves time by developing the code in a unit test and putting the tested product into the application's project. The unit test project also serves as a backup for knowing what it looked like during a good test.

Saga handler unit testing

Let's start testing saga code from the previous section in the message handler:

```
// The Test code
Test.Handler<EventMessageHandler>()
        .ExpectReply<ResponseCommand>(m => m.state ==
        PaymentMessages.MessageParts.StateCodes.
        CompleteMyWCFClient)
        .OnMessage<SendCommand>(command);
```

As we can see, the NServiceBus testing API makes use of a Fluent API specification style testing pattern as opposed to the more traditional assertion style that most people would normally use that are part of nunit or other xunit type frameworks.

We will now start testing the saga code from the `UnitTestHandlers` project and the `UnitTestSaga2.cs` file.

One thing to note is that if a saga entity object is deleted in different function calls, with the `MarkAsComplete()` method, these should be tested separately. This is because once we delete the object, we cannot delete it again. For example, in our tests, we will use the following:

```
[TestMethod]
public void Run()
{
    Test.Initialize();

    /**
     * State sent to Saga
     * ***/
    SendCommand command = new SendCommand();
    command.RequestId = new Guid("8b265223-dc9e-4789-a6df-69d19f644ad7");
    command.state = MyMessages.MessageParts.StateCodes.SentMyWCFClient;

    /**
     * Response from WCF and to MVCApp
     * ***/
    ResponseCommand resp = new ResponseCommand();
    resp.RequestId = new Guid("8b265223-dc9e-4789-a6df-69d19f644ad7");
    resp.state = MyMessages.MessageParts.StateCodes.CompleteMyWCFClient;


    Test.Saga<MyTestSaga>()
            .ExpectReplyToOrginator<ResponseCommand>()
            .ExpectSend<SendCommand>()
        .When(s => s.Handle(command))
            .ExpectReplyToOrginator<ResponseCommand>()
        .When(s => s.Handle(resp))
        .AssertSagaCompletionIs(true);

    Test.Saga<MyTestSaga>()
            .ExpectReplyToOrginator<ResponseCommand>()
            .ExpectSend<SendCommand>()
            .ExpectTimeoutToBeSetIn<SendCommand>((state, span) => span == TimeSpan.FromHours(3))
        .When(s => s.Handle(command))
            .ExpectReplyToOrginator<ResponseCommand>()
        .WhenSagaTimesOut()
        .AssertSagaCompletionIs(true);

}
```

Saga Development

In this snippet, we are testing the message handler with two separate conditions. The first test case is the normal condition of a saga start where we are testing the `IHandle Messages<ResponseCommand>` message handler.

The second test case is the timeout condition where we are testing the `IHandleTimeouts<SendCommand>` handler. These two test cases were used in the same file as they reused some of the same pieces.

The saga handler itself will act as a mediator between MVCApp and the WCF client. This is needed to act as a timeout after three hours in case there is no response from the WCF service.

Integration tests with MVC

Normally, when putting a bus in MVC, we wish to create the bus only once, and then reuse it over and over again from different controllers to send messages. In order to do this, we will be putting the code in the `Global.asax.cs` file under `..\MVCApp - WCF\MVCApp\`.

However, we will perform integration tests to ensure that all the pieces are working. Performing tests outside of the MVC application itself can assist in isolating databases and endpoints that are not deployed and are needed for the application to process. The integration test can be found in the `MVCToNSBTests.cs` file under `..\MVCApp - WCF\IntegrationTests\\`. There are two items that we have added to the MVC application: an endpoint to send messages to the saga and an MVCApp database with Entity Framework connections to store the payment messages in the `PayMessage` table.

In order to create this table, the MVCApp database must be created; in this case, `.\SQLExpress`. After the database is created, we can create the table structure by clicking on the **Generate Database from Model...** option, and then executing the SQL script that was created from this execution, as shown here:

Chapter 4

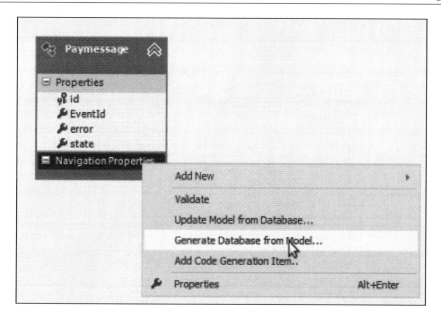

Now that the database portion has been created, we need to create the `mysaga` endpoint in MSMQ. This is done by simply executing the `MySaga` solution as follows:

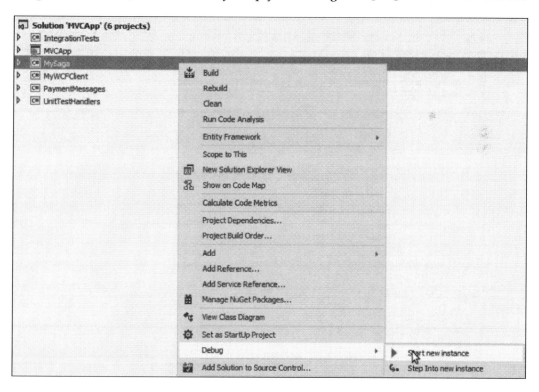

Saga Development

We can view in the **MSMQ Commander** window that the endpoint was created, as shown here:

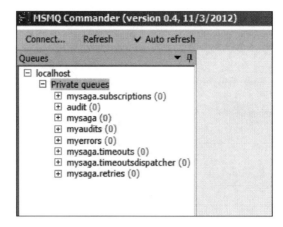

If the tests run successfully, then the `mysaga` endpoint is present and working. We can see the test program working as follows:

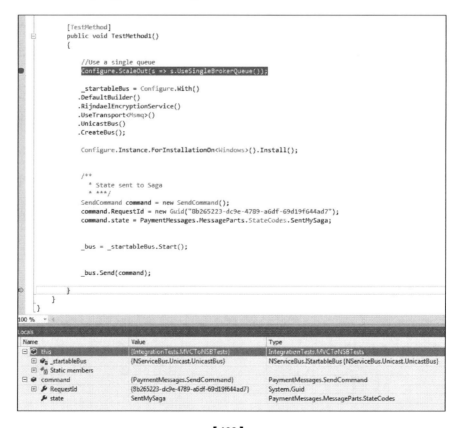

Chapter 4

We can see the test program in **Test Explorer** working as follows:

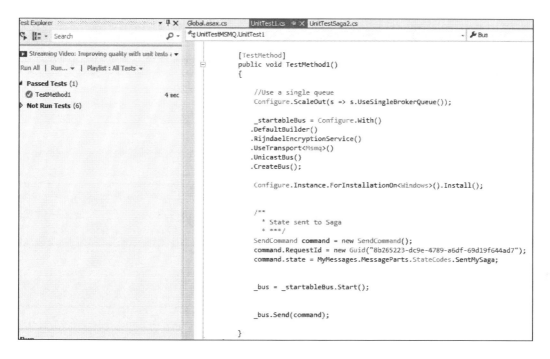

Now, we can copy this code into the `Application_Start()` function in the `Global.asax` file to be called at the startup of the MVC and use it as the bus for the MVC.

We create a controller to use this bus to send the command by ID; it can be selected from the MVC view as follows:

[133]

Saga Development

We will send the message to the saga by the methods in the MVC controller. The controller will be selected by ID, and a lookup in the table for the correct message will be made from the `PayMessage` table. The lookup will be done in an Entity Framework connecting the **data access layer (DAL)** of the code. The sending to the saga appears as follows:

```
public ActionResult SendSaga(int id)
{
    //Get the GUID from the available XML files fro, the id passed from the page
    var user = new XMLLoads().GetPayments().Where(p => p.Id == id).FirstOrDefault();
    var message = new XMLLoads().GetMessages().Where(p => p.EventId == user.EventId).FirstOrDefault();

    // Create the send
    SendCommand command = new SendCommand();
    command.RequestId = message.EventId;
    command.state = StateCodes.SentMySaga;

    // Change the state in the EF DAL
    MVCAppDAL dal = new MVCAppDAL();
    dal.changeState(command);

    // Send the command on the Bus
    MvcApplication.Bus.Send(command);

    return View(user);
}
```

The DAL will read the database table called `PayMessage`, which is configured as an entity model object from the `Model1.edmx` file that does the ORM mapping. The connection string pointing towards the table and database is defined in the `Web.config` file. The `PayMessageModel.edmx` file was generated from the database to provide the mapping to the objects. So first, we need to build a database table to contain the `PayMessage` table that looks like the following with GUIDs and state. This is used to update the state from the messages for the MVCApp.

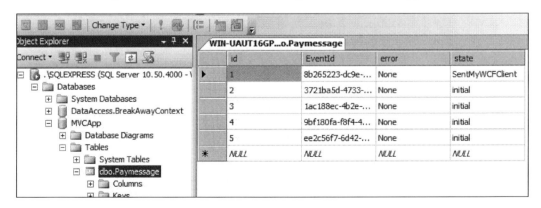

The database table, `<PayMessage>`, has the following properties, as shown here:

Column Name	Data Type	Allow Nulls
id	int	☐
EventId	uniqueidentifier	☐
error	varchar(50)	☑
state	varchar(50)	☑

RabbitMQ for NSB

RabbitMQ is a cross-platform messaging framework, like MSMQ, that can run on both Linux and Windows operating systems.

RabbitMQ has many more features than MSMQ, such as routing, virtual hosts, and a powerful admin toolset. It can run on Linux and scales out very well. For large enterprise systems (especially in heterogeneous network environments), you should really consider RabbitMQ. RabbitMQ can easily be changed for the MSMQ queuing system in the NSB configuration.

For administrating RabbitMQ, a web admin interface can easily be installed. However, if the command-line interface is required, then the Python language will need to be installed.

To revisit some of the references in the earlier chapters, take a look at the following:

- **Local host management site**: This can be found at http://localhost15672/#/.
- **Documentation**: This can be found at http://www.rabbitmq.com/documentation.html.
- **Windows installation**: This can be found at https://www.rabbitmq.com/install-windows.html.
- **NServiceBus samples**: These can be found at https://github.com/Particular/NServiceBus.RabbitMQ.Samples.
- **Development tools site for RabbitMQ**: This can be found at http://www.rabbitmq.com/devtools.html.
- **NSeviceBus.RabbitMQ source code**: This can be found at https://github.com/Particular/NServiceBus.RabbitMQ.
- **NServiceBus RabbitMQ hands-on lab**: This can be found at http://particular.net/HandsOnLabs.

After installing RabbitMQ, we can set the management plugin using the `rabbitmq plugins` to enable `rabbitmq_management`. This is to view the queues using the web management `http://localhost:15672/#/queues`, as shown in the following screenshot:

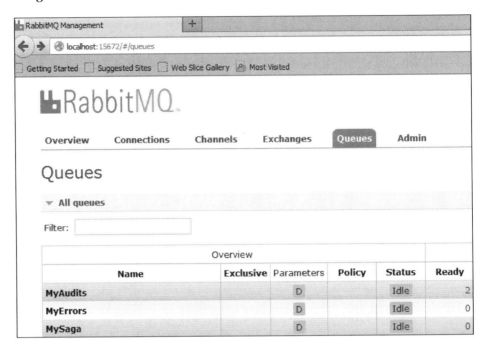

We can get a list of queues by entering `rabbitmqctl list_queues` in the RabbitMQ command prompt, as shown here:

We can delete all the queues by running the following:

```
rabbitmqctl stop_app
rabbitmqctl reset
rabbitmqctl start_app
```

The source code

The directory for the code is under the `RabbitMQ` directory. There are two solutions, which are as follows:

- `MVCApp - WCF`: This is used to send WCF messages as a client using a saga. But instead of MSMQ, RabbitMQ is used for queuing.
- `WCFService`: This is used as the WCF service.

The solution was built in VS 2012 in several operating systems, including Windows Server 2012, Windows Server 2008, and Windows 8.1, with MSMQ, DTC, RavenDB, NServiceBus Version 4.0 references, and SQL Server 2012 Express LocalDB installed.

Changing the endpoints

There are going to be subtle differences in setting up the endpoint configurations. These are the three basic steps:

1. Add the `NServiceBus.RabbitMQ` reference.
2. Change the NServiceBus transport mechanism from `<MSMQ>` to `<RabbitMQ>`.
3. Set the RabbitMQ transport configuration in the `App.config` file.

In the `MySaga` project, we will be making changes to the `App.config` and `EndpointConfig.cs` files.

Saga Development

The `NServiceBus.RabbitMQ` package will have to be installed into each project to support RabbitMQ. It will be added via NuGet. Go to http://www.nuget.org/packages/NServiceBus.RabbitMQ/ for more information.

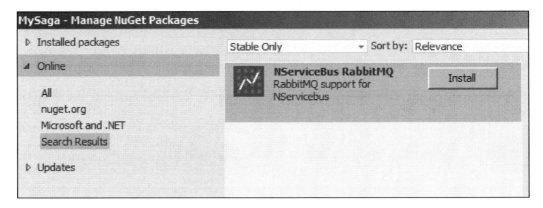

The `.UseTransport<>` method that is defaulted to MSMQ will have to be switched to RabbitMQ, as shown here:

```
Configure.With()
    .DefaultBuilder()  // Autofac Default Container
    .UseTransport<NServiceBus.RabbitMQ>()
    .InMemorySubscriptionStorage()
    .UseNHibernateSagaPersister()
    .UseNHibernateTimeoutPersister()
    .UnicastBus(); // Create the default unicast Bus
```

We will have to set the transport mechanism to the local host as this is where the RabbitMQ service is residing. This is shown in the screenshot here:

```xml
  <!--<section name="AuditConfig" type="NServiceBus.Config.AuditConfig, NServiceBus.Core" />-->
</configSections>
<!-- NHibernate Settings-->
<connectionStrings>
  <add name="NServiceBus/Transport" connectionString="host=localhost" />
  <add name="NServiceBus/Persistence" connectionString="Data Source=.\SQLEXPRESS;Initial Catalog=nservicebus;
</connectionStrings>
<!-- specify the other needed NHibernate settings like below in appSettings:-->
```

Beyond the changes discussed, there are only very little changes needed to move to different queuing systems. The saga and message handlers work in the same way; we are only changing the endpoint transportation mechanisms.

We can see that queues were created and run from this example in a previous screen from the queues screenshot, which is shown at http://localhost:15672/#/queues.

ActiveMQ in NSB

Apache Active Message Queue (ActiveMQ) is a JAVA open source framework from the Apache foundation based on the **Java Message Service (JMS)**. Visit http://en.wikipedia.org/wiki/Apache_ActiveMQ and https://activemq.apache.org for more information. It will run on a machine, be it Windows or Linux, in a **Java Runtime Environment (JRE)**. JAVA has to be operational on the machine and have the environment path for JAVA_HOME configured to point at the root folder of the JRE. The installation instructions for ActiveMQ can be found at https://activemq.apache.org/getting-started#GettingStarted%20-Download.

The source code

In this section, we will be using the ActiveMQ solution. This solution is similar to RabbitMQ, except ActiveMQ is used instead of RabbitMQ.

There will be three basic steps:

1. Add the NServiceBuActiveMQ reference.
2. Change the NServiceBus transport mechanism from <MSMQ> to <ActiveMQ>.
3. Set the ActiveMQ transport configuration in the App.config file.

Once downloaded on the Windows OS, we unzipped the Window's binary files into the c:\activemq\ directory. Running the activemq.bat batch file from the command prompt, in c:\activemq\bin\, will display a series of commands to show that the ActiveMQ is running. This is shown in the following screenshot:

Saga Development

An alternative is to install ActiveMQ as a Windows service in which installation scripts exist for both Win32 and Win64 machines. For a 64-bit Windows Server, we can use `InstallService.bat` in `c:\activemq\bin\win64\`.

Ensure that RabbitMQ is not running as a Windows Service in the background, as they utilize the same network ports. Also, Microsoft ServiceBus for Windows Servers will share the same JMX ports as well, which will be port 5672. ActiveMQ's default port is 61616. Checking the ports can be done with Micrsoft's TcpView, which can be found at `http://technet.microsoft.com/en-us/sysinternals/bb897437`.

To ensure that ActiveMQ is running, you may access the admin console in the browser by using `http://localhost:8161/admin`. The default user ID and password are `admin` and `admin` respectively. Please visit `http://activemq.apache.org/getting-started.html` to have a look at the documentation. When accessing the admin console, you should get something that looks like the screenshot here:

To use ActiveMQ for NServiceBus in Visual Studio projects, the NuGet version of `NServiceBus.ActiveMQ` has to be installed. Go to https://www.nuget.org/packages/NServiceBus.ActiveMQ/1.0.5 to look at installing `PM> Install-Package NServiceBus.ActiveMQ` into the projects. Ensure that the configuration for the IBus is set for ActiveMQ, as shown here:

```csharp
protected void Application_Start()
{
    Configure.ScaleOut(s => s.UseSingleBrokerQueue());
    _startableBus = Configure.With()
    .DefaultBuilder()
    .RijndaelEncryptionService()
    .UseTransport<ActiveMQ>()
    .UnicastBus()
    .CreateBus();
    Configure.Instance.ForInstallationOn<Windows>().Install();

    _bus = _startableBus.Start();
    AreaRegistration.RegisterAllAreas();

    RegisterGlobalFilters(GlobalFilters.Filters);
    RegisterRoutes(RouteTable.Routes);
}
```

Ensure that the `App.Config` or `Web.Config` file have the appropriate connection string for the `NServiceBus/Transport` string to point at the correct instance of the ActiveMQ queues:

```xml
<connectionStrings>
<add name="NServiceBus/Transport"
  connectionString="ServerUrl=activemq:tcp://localhost:61616"/>
</connectionStrings>
```

Saga Development

Some of the `EndpointConfig.cs` files may not explicitly call the `IBus.Configure()` method. So make sure that the ActiveMQ using transport call is explicitly called in the `EndpointConfig` class:

```
namespace MySFTPClient
{
    using NServiceBus;

    /*
        This class configures this endpoint as a Server. More information about how to configure the NServiceBus host
        can be found here: http://particular.net/articles/the-nservicebus-host
    */
    public class EndpointConfig : IConfigureThisEndpoint, AsA_Publisher, UsingTransport<ActiveMQ>
    {
    }
}
```

When we start the MVC application and associated saga code, we can see that the queues are created in ActiveMQ by looking through the admin console of ActiveMQ at `http://localhost:8161/admin/queues.jsp`. Notice that the queues were created matching these programs. When we execute this example, we should have queues created in the ActiveMQ administration tool as shown here:

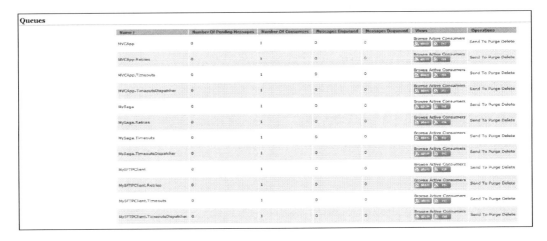

At this point, we see that the queues are working for ActiveMQ, and we have a program that we can now start extending to use ActiveMQ with an MVC frontend, using Entity Frameworks, into saga data and user tables. This example will be the ActiveMQ solution.

Chapter 4

There were not many changes needed in the code to change it from the RabbitMQ queues to the ActiveMQ queues. We can see the queuing in the Windows consoles as they run in the screenshot here:

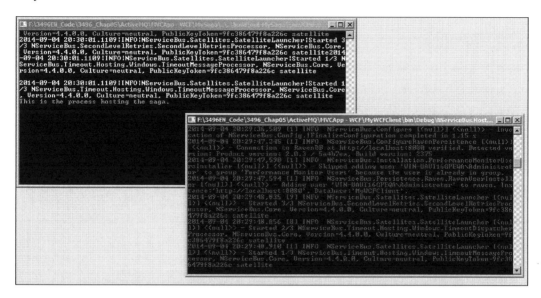

Summary

In this chapter, we have discussed about building a WCF server application. Then we looked at building an MVC application that interfaces into a WCF client to communicate with the WCF server.

We extended the MVC-WCF example by constructing sagas and message handlers in a unit testing environment. We then moved onto led into unit testing sagas and message handlers as we discussed testing using the `NServiceBus.Testing` framework.

We took a deep dive into our example after adding WCF, and later added the bus to decouple the browser from services like WCF so as to enable the user to continue in the browser. This chapter also took into account the errors that could occur in web services.

We then took the example and changed the transport mechanism from MSMQ to RabbitMQ with minor changes. We discussed in testing how we can change sagas and message handlers to be similar, and also discussed how we can change them enough to enhance their use. We also briefly discussed the many NServiceBus testing rules to build the handlers in the unit testing environment, without worrying about the endpoints until later.

We will continue in the upcoming chapters to go through more snippets and scenarios. We will also go into greater detail about using handlers for transactional and error handling needs. We will also talk further about using tools in these upcoming discussions.

5
Saga Snippets

In this chapter, we will be focusing on snippets in sagas. We will discuss an e-mail and **Secure File Transfer Protocol (SFTP)** example that will be set to timeout by a daily timer in saga code. The saga code will be a mediator between a frontend **Windows Presentation Framework (WPF)** and a backend client executing either e-mail or SFTP. Using a saga as a mediator between frontend and backend code that will interface into an external server will offer many added benefits and features. The external server interface, such as an e-mail server or SFTP server, is usually beyond our control and is in the control of external operations or organizations, such as a bank. So, the interface into these servers is all that we have to work with, and as business, software, and operational needs increase, we need a framework robust enough to meet these demands. Thus, we have NSB and sagas.

We will also walk through changing this application to support ActiveMQ. We will briefly discuss ActiveMQ and how to set it up to perform these operations.

In this chapter, we will cover:

- Sample e-mail saga notification
- Sample SFTP saga
- Saga deployment
- ActiveMQ

Saga Snippets

Source code overview

In this chapter, there will be three directories of source code:

- A directory of the solution will be found in the `EmailSagaTest` directory. This program is a timer-based program using an NSB saga to send a daily e-mail containing information regarding MSMQ. This solution was designed to send a daily e-mail to operations of a system status.
- A directory of the solution will be found in the `SFTPSagaTest` directory. This program is a timer-based program using an NSB saga to send SFTP files to an SFTP server. This solution was designed to send files to banks.
- A directory of the solution found will be in the MVCApp - ActiveMQ directory.

This program shows the use of ActiveMQ.

All source code was built with Visual Studio 2012, and used in Windows 8.1, Windows Server 2008, and Windows Server 2012. All programs were built with NServiceBus and require NServiceBus to be installed using MSMQ.

Sample e-mail saga notification

We mentioned earlier that normal production is filled with notifications checking queues, tables, processes, tasks, and more. We will create an NSB saga program that is based on a timeout to send a message to operations or to ourselves as developers.

The frontend of the application will use **Windows Presentation Framework (WPF)**, using the **Extensible Application Markup Language (XAML)**. An introduction to XAML can be found at `http://msdn.microsoft.com/en-us/library/ms752059(v=vs.110).aspx`.

The controller piece in the middle of the application is used to apply the timeout, route the messaging, and persist the state in the saga. The piece that the controller will communicate with to send the e-mail is the email client, which will have a message handler to communicate with the saga. This interaction has been depicted in the following diagram:

Chapter 5

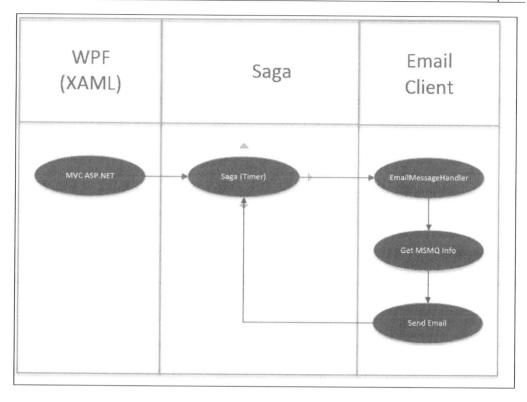

The requirement of this application is that we set a timer through a GUI. This GUI will also allow some e-mail variables to be set as well, such as the email server, and the To and From addresses. The timer will set a time in the saga to send a daily e-mail. The daily e-mail will be a simple e-mail to send to operations the available MSMQ queues. We do not wish to hardcode the timer or any of the e-mail variables. We also do not wish to use the Windows Task Manager as that will require more operational support. This exercise will be handled easily in NSB.

Separating the frontend GUI from the backend email client through a saga that is acting as a persistence and message mediator provides several benefits:

- **Separation of duties**: If the frontend has issues, the saga does not necessarily propagate those issues to the email client, as long as the messaging is correct.
- **Message durability**: Depending on the configuration, the messages, endpoints, and saga data is persisted to the point where the server can crash, but the data is still recoverable.

- **Retries**: If a message fails, it will retry several times based on the configuration.
- **Monitoring**: There are many tools and depending on the persistence and queuing configurations, many ways to check the endpoints, messages, and services.
- **High availability**: NSB is a high-availability framework, meaning that there are multiple connections, endpoints, and server and messaging scenarios for configuration to ensure that services are always running in the background and receiving information in a high-performance environment.

Many of these topics have been covered, but as requirements are discussed for this sample application, these features will be added by just using NSB.

Using XAML

XAML is a declarative XML-based language developed by Microsoft that is used for initializing structured values and objects for graphical presentation. It is particular to .NET. By using XAML, you can separate the graphical designer code created in XAML from the code that defines logic created in C#. For instance, in XAML, a button can be designed, and in C#, the logic of what happens when the button is clicked can be decided.

We will be extending the WPF sample of a timer found at `http://www.codeproject.com/Articles/237011/CREATING-A-CUSTOM-TIMEPICKER-CONTROL-IN-WPF`. Not only can XAML be built with the toolbox that is enclosed with Visual Studio in a visual designer, but XAML interfaces are designed to be graphically built in Microsoft Blend as well.

Microsoft Blend for Visual Studio is a Microsoft-developed user interface design tool for creating interfaces for both, the web and the desktop. It is meant to blend the two types of applications. Here's a look at the timer test example in Blend:

Chapter 5

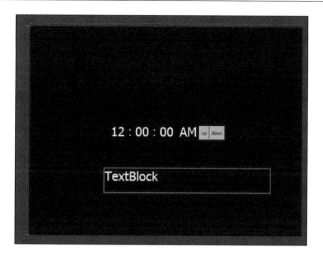

Microsoft Blend has a wide range of tools for building user interfaces in a **What You See Is What You Get (WYSIWYG)** visual editor, an example of which has been provided in the following screenshot:

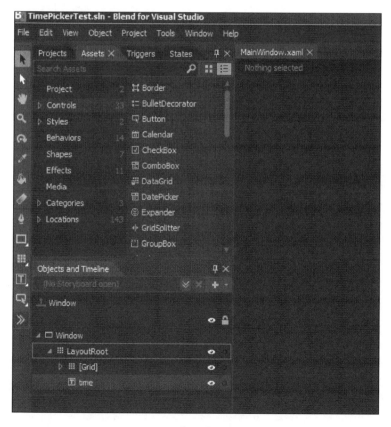

Saga Snippets

While these are some of the benefits of using XAML and WPF, for our simple purposes, in the `TimerPickerTest` project, we will be using the toolbox that is enclosed with Visual Studio 2012 to do our graphics to match the following:

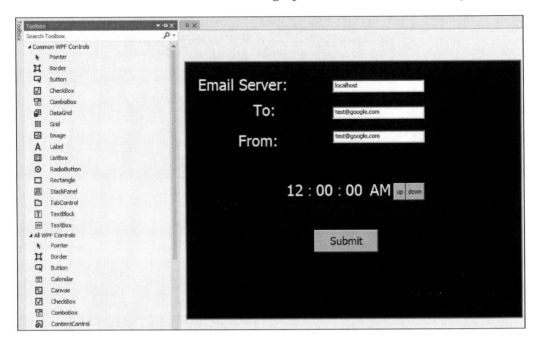

This GUI is to set several variables: the Email Server, which by default is set as the localhost, and the `To` and `From` addresses, which are both set by default as `test@google.com`.

The saga project

The saga project in this solution will be called `TimerSaga`. It will be dependent on the messages found in the `TimerMessages` project that will contain the available NSB messages.

This project will look like the typical saga projects that we have built thus far, as follows:

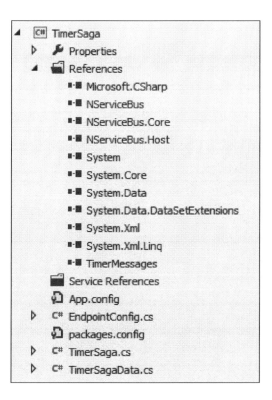

We see that we are referencing the NServiceBus.Host reference, which in turn created the EndpointConfig.cs file to contain the code for the bus configuration. We also know that it must have set the NServiceBus.Host.exe file to debug this project as a DLL through NSB. An App.config file was also created with the NSB host reference to include the unicast and other settings for the EndpointConfig.cs file.

The App.config file will contain the setting to send SendTime, the message with timing information, and EmailMessage, the message with the e-mail information, to the EmailClient endpoint, which will have a message handler to send the operational e-mail. This setting will be as follows:

```
<UnicastBusConfig>
  <MessageEndpointMappings>
    <add Endpoint="EmailClient" Messages="TimerMessages.EmailMessage, TimerMessages" />
    <add Endpoint="EmailClient" Messages="TimerMessages.SendTime, TimerMessages" />
  </MessageEndpointMappings>
</UnicastBusConfig>
```

Saga Snippets

The saga will mostly be made of the saga class, with the various sagas and message handler functions, and the saga data. The saga class will be called `TimerSaga`. The saga data class will be called `TimerSagaData`. The `TimerSagaData` class will store saga unique information and the timer information normally in the form of hour text, minute text, and second text. The data will also store e-mail information, such as details of the email server and the `To` and `From` address information.

The `TimerSaga` class will have its starting message, in this case, `SendTime`. There will be two message handlers: the `ResponseCommand` from the `EmailClient`, and the `TimeoutMessage` for when the timer is timed out. The `setDateTime` function will calculate if we are looking at the timer for today or tomorrow. When the timeout is called, it will need to set the timer for tomorrow as well. The diagram of `TimerSaga` class appears as follows:

Chapter 5

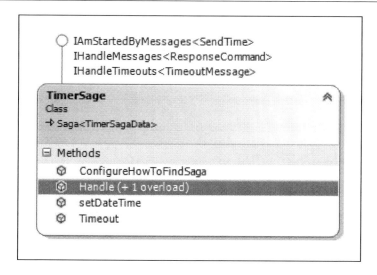

When `TimerSaga` will receive its starting message—in this case, `SendTime`—it will save the saga data and set the timer. It sends a message to the e-mail handler just to check if it is functional. The code will appear as follows:

```
/*
 * SendTime Handler
 * */
public void Handle(SendTime message)
{
    Data.RequestId = message.RequestId;

    TimeoutMessage tmMessage = new TimeoutMessage();
    tmMessage.RequestId = message.RequestId;

    DateTime toSet = (DateTime)setDateTime(message);

    /**
     * Save Email Info
     * */
    Data.emailServer = message.emailServer;
    Data.toAddress = message.toAddress;
    Data.fromAddress = message.fromAddress;
    /**
     * Save Timer Info
     * */
    Data.HrTxt = message.HrTxt;
    Data.MinTxt = message.MinTxt;
```

Saga Snippets

```
        Data.SecTxt = message.SecTxt;
        Data.AmPmTxt = message.AmPmTxt;


        RequestTimeout(toSet, tmMessage);
        Bus.Send(message);

}
```

When the saga times out after the daily time is hit, it will send the e-mail information to the e-mail handler to process the e-mail, and set the timer for the next execution as follows:

```
        /*
         * Timeout
         * */
        public void Timeout(TimeoutMessage message)
        {

            /**
             * Retrieve Email Info
             * */

            EmailMessage emailMessage = new EmailMessage();
            emailMessage.emailServer = Data.emailServer;
            emailMessage.fromAddress = Data.fromAddress;
            emailMessage.toAddress = Data.toAddress;
            /**
              * Get Timer Info
              * */
            SendTime timeMessage = new SendTime();
            timeMessage.AmPmTxt = Data.AmPmTxt;
            timeMessage.HrTxt = Data.HrTxt;
            timeMessage.MinTxt = Data.MinTxt;
            timeMessage.SecTxt = Data.SecTxt;

            /**
               *
               * Reset the Timer
               *
               **/
            TimeoutMessage tmMessage = new TimeoutMessage();
            DateTime toSet = (DateTime)setDateTime(timeMessage);
```

```
            RequestTimeout(toSet, tmMessage);

            Bus.Send(emailMessage);
    }
```

Testing the solution

Before testing the application from the WPF GUI, we will create a console application to send different parameters through the message to the saga to see if it behaves as expected. This console application will be found under the `ConsoleEmailSagaTest` directory. This code will appear as follows for sending the saga a sample e-mail and timer information, where the timer is set for `7:50:00 PM`:

```
class Program
{

    private static IBus _bus;

    static void Main(string[] args)
    {

        Configure.ScaleOut(s => s.UseSingleBrokerQueue());
        _bus = Configure.With()
                    .DefaultBuilder()
                      .UseTransport<Msmq>()
                   .UnicastBus()
                    .CreateBus();


        Configure.Instance.ForInstallationOn<Windows>().Install();

        SendTime s_Time = new SendTime();
        /*
         * Email Info
         * */
        s_Time.toAddress = "test@google.com";
        s_Time.fromAddress = "test@google.com";
        s_Time.emailServer = "localhost";
        /*
         * Timer Info
         * */
        s_Time.HrTxt = "7";
        s_Time.MinTxt = "50";
```

```
            s_Time.SecTxt = "00";
            s_Time.AmPmTxt = "PM";

            _bus.Send(s_Time);
        }
    }
```

To test the e-mail, here is a **Simple Mail Transfer Protocol (SMTP)** listener that will intercept the e-mails locally on port 25 for viewing, to test your e-mail sending scenarios. It can be found at `http://smtp4dev.codeplex.com/`. When the e-mail is sent to the localhost, it will be recorded for review in the smtp4dev software.

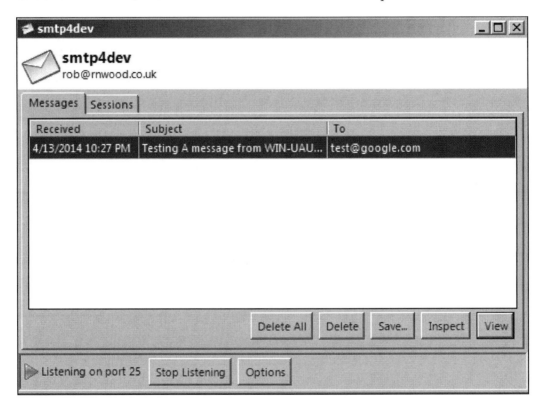

If all works well and we receive our message from the email client, we should be getting an e-mail that looks similar to the following screenshot:

We just walked through a simple solution using a GUI to configure an email client to send us a daily operational e-mail about MSMQ. This example could be easily extended to provide a daily report of SQL Server tables, the number of messages in an audit or error queue, and many other items to check the operational status of a system.

Another daily example, especially for anyone dealing with payment engines sending files to banks or for third-party transmissions, is the use of SFTP to send files. With the usage of SFTP, e-mail, and WCF in the previous chapter, many enterprise scenarios can be solved.

Saga Snippets

Sample SFTP saga

In many enterprises, you may find the need to use **SSH File Transfer Protocol** (**SFTP**) to move files securely into remote environments or even to share files across multiple clients using cloud file storage systems. SFTP is a means to upload and download files across a secure network pipe. For further information, see `http://en.wikipedia.org/wiki/SSH_File_Transfer_Protocol`.

In this section, we will be discussing the code found in the `SFTPSagaTest` directory. The requirements will be similar to the previous example on the e-mail timer; except now we will be sending a file through SFTP.

We will use the same basic design, where the frontend GUI will be WPF XAML. It will call `TimerSaga` to set the timer. `TimerSaga` will send the `SFTPMessage` to the SFTP client, which will handle the message by the `SFTPMessageHandler` to establish an SFTP connection and upload a file to the SFTP server. It will work as in the following diagram:

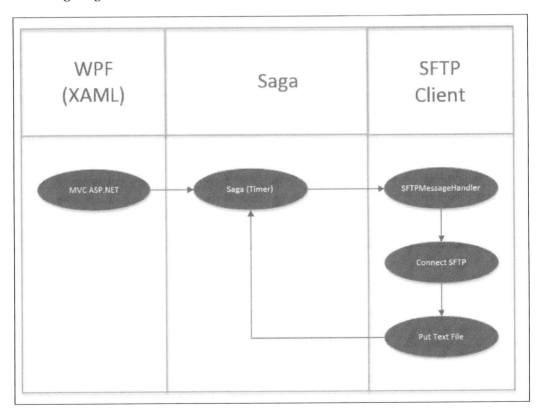

The requirement of this application is that we set a timer through a GUI. This GUI will also allow some SFTP variables to be set, such as the SFTP server, username, password, and local text file location.

The timer will set a daily time in the saga to upload a daily text file, which will be given a unique name based on date and time into the SFTP server. This avoids overwriting the same filename.

The daily upload will be a simple example of uploading a file using saga interaction to a remote SFTP server, such as a bank, for batch payments. We do not wish to hardcode the timer or some of the SFTP variables. We also do *not* wish to use the Windows Task Manager as that will require more operational support. This exercise will be handled easily in NSB.

Separating the frontend GUI from the backend email client through a saga that is acting as a persistence and message mediator offers several benefits:

- **Separation of duties**: If the frontend has issues, it does not necessarily propagate those issues to the email client, as long as the messaging is correct.

  SFTP may have issues logging in on occasion due to password resets, or network or even server issues. By not having the frontend directly tied to the SFTP client, we can avoid the frontend crashing on SFTP issues.

- **Message durability**: Depending on the configuration, the messages, endpoints, and saga data are persisted to the point that the server can crash, but the data is still recoverable.

- **Retries**: If a message fails, it will retry several times based on the configuration.

  This may allow us to try the SFTP client-to-server connections multiple times if there are errors within the SFTP connection.

- **Monitoring**: There are many tools, and depending on the persistence and queuing configurations, many ways to check the endpoints, messages, and services.

  This may allow us to monitor the SFTP client as well as the information being passed into it to validate if it is correct for an SFTP connection.

- **High availability**: NSB is a high-availability framework, meaning that there are multiple connections, endpoints, and server and messaging scenarios for configuration to ensure that services are always running in the background and receiving information in a high-performance environment.

Many of these topics have been covered, but as different requirements are discussed for this sample application, these features will be added by just using NSB.

Using XAML

We will be using XAML as a declarative XML-based language that we had developed in the previous e-mail example. The difference is that we incorporated some SFTP fields instead of e-mail fields to provide the SFTP server, username, password, and local text filename as variables, depicted as follows:

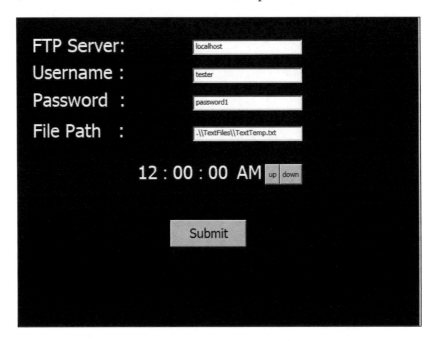

Changing the process of messaging

In the e-mail example, we used the values in `EmailMessage` to pass the information from the saga to the email client. It looked like the following diagram:

For the SFTP, we will be using `SFTPMessage` instead with SFTP variables for uploading a file as shown in the following diagram:

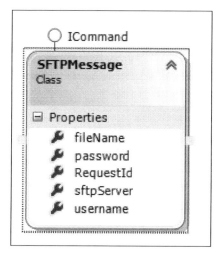

Notice that the variable matches almost exactly with the input variables for the SFTP portion of the frontend GUI. Code can be reused when creating services. In this sample, we also use the message ID, called `RequestId`, as a primary key to look up for the saga data.

The saga code will function in a similar manner to the email timer saga and email message handler to execute the email client. However, now we will be passing SFTP information from the frontend to the saga and the message handler. Instead of accessing the email client, we will be executing an SFTP client. Despite these message changes and using SFTP instead of e-mail, much of the code will remain the same.

Setting up an SFTP test environment

For testing, we will need an SFTP server and client. We have developed an SFTP client in the saga code, but we need an SFTP server to establish the connection and upload the file.

We will start by setting up a test SFTP server from http://www.freesshd.com. After simply running the installation, we will set up a user and the default directory into which the user can upload and download the files.

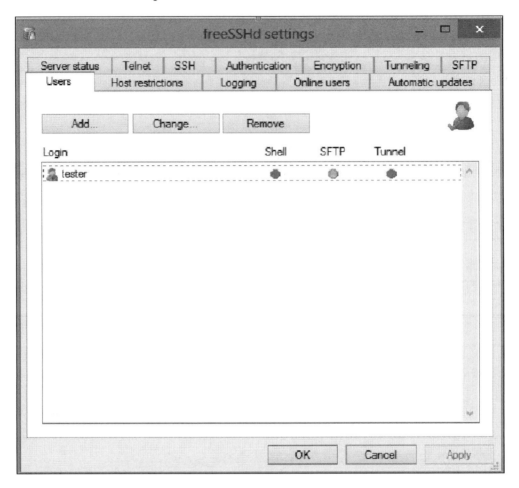

Before testing the entire solution that we have created using a saga, let's perform a simple test to ensure that the connectivity for the SFTP server is set correctly. We will write a simple console program to test out this connectivity and to see the code to put a file on the SFTP server. This sample code can be found in the `ConsoleTestSFTP` directory and will look like the following:

```
namespace ConsoleTestSFTP
{
    class Program
    {
        static void Main(string[] args)
```

```
        {
            /**
             * Connect to the SFTP
             * and put file
             * */
            try
            {
                Sftp Sftpclient = new Sftp("localhost", "tester",
                    "password1");
                Sftpclient.Connect();
                string newFileName = string.Format("text-{
                    0:yyyy-MM-dd_hh-mm-ss-tt}.txt",
                        DateTime.Now);
                Sftpclient.Put(".\\TextFiles\\TextTemp.txt",
                    newFileName);
            }
            catch (Exception ex)
            {
                Console.Out.WriteLine(ex.Message);
            }

        }
    }
}
```

In order for this code to work, we need to add SFTP framework references. We will include the `DiffieHellman` reference, the `Org.Mentalis.Security` reference, and the `Tamir.SharpSSH` reference. We can see them appear in the following screenshot:

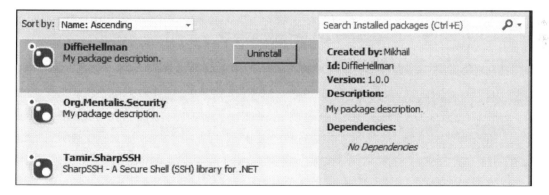

When running the sample SFTP client program, we should get no exception message. However, we should also check that the file has been uploaded correctly on the SFTP server. In order to do that, we need to look into a couple of items in freeSSHd. The first is to see where the upload directory has been set, and check the local server directory for the presence of the file. We will look in the freeSSHd GUI for the local directory as follows:

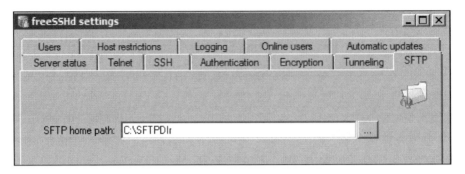

We can look in the associated \SFTPDir directory under C: and see that the file was uploaded at the time we had tested:

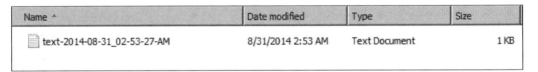

An alternative to checking if the file is uploaded via freeSSHd by looking at the file itself is to check the freeSSHD logs. We can check the **Logging** section of the freeSSHd GUI to see where the log resides or even click on the **Open** button to review it from the freeSSHd itself, as depicted in the following screenshot:

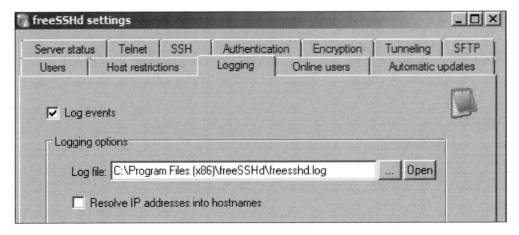

Viewing the log, we can see that the file was uploaded successfully with more details:

```
freesshd - Notepad
File Edit Format View Help
08-31-2014 02:53:26 IP 127.0.0.1 SSH connection attempt.
08-31-2014 02:53:27 IP 127.0.0.1 SSH tester successfully logged on using password.
08-31-2014 02:53:27 SFTP service granted to user tester.
08-31-2014 02:53:29 IP 127.0.0.1 tester is uploading /text-2014-08-31_02-53-27-AM.txt (C:\SFTPDir\text-2014-08-31_02-53-27-AM.txt)
08-31-2014 02:53:41 IP 127.0.0.1 SSH tester disconnected.
```

Saga deployment

In the SFTP and e-mail examples, we executed many of the pieces from the Visual Studio solutions as DLLs. NServiceBus hosts these assemblies and with its underlying pipeline will instantiate and execute a specific module. NServiceBus hosts these projects when installing the NServiceBus.Host reference; in turn, it will execute the associated DLL with the NServiceBus.Host.exe executable.

These DLLs are run by the executable NServiceBus.Host.exe. The NServiceBus.Host.exe file is a utility program to not only run NServiceBus host programs in a console program, but also install the DLLs as Windows Services. This executable makes use of the TopShelf framework, and just as TopShelf has code configurations for Windows Services, so does NServiceBus.Host.exe.

Moreover, it is TopShelf capability to run a console program as a console application while debugging, and run it as a Windows service when the program is installed. This is the main reason TopShelf is so popular; otherwise, if you create a Windows service using .NET and Visual Studio, you have to create separate projects to debug or to run as a Windows service.

NServiceBus installers are effective ways to plug in your bootstrapping code like creating queues, folders, databases, and so on. The following code shows how to invoke the installers manually:

```
_bus = Configure.With()
    .DefaultBuilder()
    .UseTransport<Msmq>()
    .UnicastBus()
    .CreateBus();

    Configure.Instance.ForInstallationOn<Windows>().Install();
See http://docs.particular.net/NServiceBus/
    nservicebus-installers
```

Saga Snippets

This can be done by adding the NServiceBus configure `.Start(() => Configure.Instance.ForInstallationOn<NServiceBus.Installation.Environment.Windows>().Install());` in code or adding it in the `NServiceBus.Host.exe \ install` script. This is commonly known as self-hosting.

In order to write your own installer and manually configure it with NServiceBus, just implement a class which implements the `INeedToInstallSomething<T>` interface as shown in the following code:

```
public interface INeedToInstallSomething<T> : IneedToInstallSomething
where T: Ienvironment
{
  void Install(string identity);
}
```

We can run `NServiceBus.Host.exe MySFTPClient.dll` from the command prompt to start the host program.

Or we can install the service using `NServiceBus.Host.exe /install MySFTPClient.dll`, and uninstall the service using `NServiceBus.Host.exe /uninstall MySFTPClient.dll`.

Chapter 5

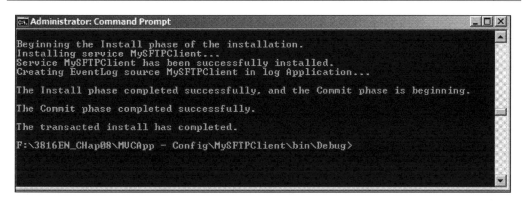

To find out more on the available commands that can be scripted from NServiceBus.Host.exe, please see http://docs.particular.net/NServiceBus/the-nservicebus-host/, or run NServiceBus.Host.exe.

We currently have the choice to have Windows services always running and to execute NServiceBus.exe based on a set time defined in the message to the saga.

An alternative is to use the Windows Task Scheduler to run a console application every 10 minutes, hourly, daily, monthly, and more. However, SFTP files could be sent to the server between 10 AM until 4 PM, for instance, based on a precondition that a message is sent to verify that the SFTP file has been validated. Business requirements can be based upon a variety of conditions, and by using NSB directly, we have taken many possible requirements into account to easily add enhancements, such as encryption without rewriting the original functionality.

Another piece of advice is to explore the Quartz framework which exists in both C# and Java, which is used to schedule cron jobs in code. Credit goes to Mark Huber, who is the author of TopShelf. Quartz integrates Quartz into the TopShelf DSL.

The C# version can be found at http://www.quartz-scheduler.net/.

[167]

In the following snippet, a service is constantly running but it runs a scheduled job at 9 AM every morning:

```
public void Start()
{
    _log.Info("Service is Started");

    Console.WriteLine("Service is Started");
    try
    {
        if (_log.IsDebugEnabled) _log.Debug("NLog successfully initialized.");
        IJobDetail jobDetail = JobBuilder.Create<CronJob>() .WithIdentity("SampleJob", "SampleJobGroup").Build();

        _log.Debug("cronTrigger started.");
        ICronTrigger cronTrigger = (ICronTrigger)TriggerBuilder.Create()
                                    .WithIdentity("SampleTrigger", "SampleTriggerGroup")
                                    .WithCronSchedule("0 0 9 1/1 * ? *")
                                    .Build();


        var schedulerFactory = new StdSchedulerFactory();
        var scheduler = schedulerFactory.GetScheduler();
        scheduler.ScheduleJob(jobDetail, cronTrigger);
        scheduler.Start();
```

The cron job is scheduled using a notation for 9 AM as `0 0 9 1/1 * ? *`. To understand the shortcut of cron scheduled notation, there are many websites that can help, such as `http://www.cronmaker.com/`. However, we must be careful here, as Quartz adds the unit of seconds making it slightly different from standard cron.

ActiveMQ

Apache Active Message Queue (ActiveMQ) is a Java open source framework from the Apache foundation based on the **Java Message Service (JMS)**. See `http://en.wikipedia.org/wiki/Apache_ActiveMQ` and `https://activemq.apache.org` for more information. It will run on a machine, be it Windows or Linux, in a **Java Runtime Environment (JRE)**. Java has to be operational on the machine and have the environment path for the `JAVA_HOME` configured to point at the root folder of the JRE. The installation instructions for ActiveMQ can be found at `https://activemq.apache.org/getting-started#GettingStarted%20-Download`.

The source code

In this section, we will be using the ActiveMQ solution; this solution is similar to RabbitMQ, except ActiveMQ is used instead of RabbitMQ.

Once downloaded on the Windows OS, unzip the Window's binary files into the `c:\activemq\` directory. Running the `activemq.bat` batch file from the command prompt, having the path as `c:\activemq\bin\`, will display a series of commands to show that the ActiveMQ is running.

Chapter 5

Ensure that RabbitMQ is not running as a Windows service in the background as it utilizes the same network ports as ActiveMQ. To ensure that ActiveMQ is running, you may access the admin console in the browser by using `http://localhost:8161/admin`. The default user ID and password is `admin` and `admin` respectively. Please visit `http://activemq.apache.org/getting-started.html` for more information. It will appear as in the following screenshot:

Saga Snippets

To use ActiveMQ for NServiceBus in Visual Studio projects, the NuGet version of `NServiceBus.ActiveMQ` has to be installed. Visit https://www.nuget.org/packages/NServiceBus.ActiveMQ/1.0.5 on installing `PM> Install-Package NServiceBus.ActiveMQ` into the projects. Ensure that the configuration for the IBus is set to ActiveMQ.

```
protected void Application_Start()
{

    Configure.ScaleOut(s => s.UseSingleBrokerQueue());
    _startableBus = Configure.With()
    .DefaultBuilder()
    .RijndaelEncryptionService()
    .UseTransport<ActiveMQ>()
    .UnicastBus()
    .CreateBus();
    Configure.Instance.ForInstallationOn<Windows>().Install();

    _bus = _startableBus.Start();
    AreaRegistration.RegisterAllAreas();

    RegisterGlobalFilters(GlobalFilters.Filters);
    RegisterRoutes(RouteTable.Routes);
}
```

Ensure that the `App.Config` or `Web.Config` file have the appropriate connection string for `NServiceBus/Transport` to point at the correct instance of the ActiveMQ queues.

```
<connectionStrings>
<add name="NServiceBus/Transport" connectionString="ServerUrl=activemq
:tcp://localhost:61616"/>
</connectionStrings>
```

Some of the `EndpointConfig.cs` files may not explicitly call the `IBus.Configure()` method, so make sure that the ActiveMQ using the transport call is explicitly called in the `EndpointConfig` class.

```
namespace MySFTPClient
{
    using NServiceBus;

    /*
        This class configures this endpoint as a Server. More information about how to configure the NServiceBus host
        can be found here: http://particular.net/articles/the-nservicebus-host
    */
    public class EndpointConfig : IConfigureThisEndpoint, AsA_Publisher, UsingTransport<ActiveMQ>
    {
    }
}
```

When we start the MVC application and associated saga code, we can see that the queues are created in ActiveMQ by looking through the admin console of ActiveMQ at `http://localhost:8161/admin/queues.jsp`. Notice that the queues were created matching these programs. When we execute this example, we should have queues created in the ActiveMQ administration tool as follows:

Queues						
Name	Number Of Pending Messages	Number Of Consumers	Messages Enqueued	Messages Dequeued	Views	Operations
MVCApp	0	1	0	0	Browse Active Consumers	Send To Purge Delete
MVCApp.Retries	0	1	0	0	Browse Active Consumers	Send To Purge Delete
MVCApp.Timeouts	0	1	0	0	Browse Active Consumers	Send To Purge Delete
MVCApp.TimeoutsDispatcher	0	1	0	0	Browse Active Consumers	Send To Purge Delete
MySaga	0	1	0	0	Browse Active Consumers	Send To Purge Delete
MySaga.Retries	0	1	0	0	Browse Active Consumers	Send To Purge Delete
MySaga.Timeouts	0	1	0	0	Browse Active Consumers	Send To Purge Delete
MySaga.TimeoutsDispatcher	0	1	0	0	Browse Active Consumers	Send To Purge Delete
MySFTPClient	0	1	0	0	Browse Active Consumers	Send To Purge Delete
MySFTPClient.Retries	0	1	0	0	Browse Active Consumers	Send To Purge Delete
MySFTPClient.Timeouts	0	1	0	0	Browse Active Consumers	Send To Purge Delete
MySFTPClient.TimeoutsDispatcher	0	1	0	0	Browse Active Consumers	Send To Purge Delete

At this point, we see that queues are working for ActiveMQ and we have a program that we can now start extending to use ActiveMQ with an MVC frontend using Entity Frameworks into saga data and user tables. This example will be MVCApp – ActiveMQ.

Summary

In this chapter, we took a deeper dive into an example that uses an email client to send a message to an email server, and used the NServiceBus saga to handle a lot of the message routing. We extended the example to use an SFTP client as well to test into an SFTP server.

We discussed many associated pieces that can assist working with enterprise development and sagas, which include checking the queues and sending notices through an e-mail, as well as checking the status of files on drives. While this chapter may be using older technologies, such as e-mail, file operations, and SFTP, there are many environments that rely on e-mails for notifications, SFTP for sending and receiving files, and file operations throughout the environment. C# provides many frameworks and greater ability to access the lower-level APIs for logging and monitoring.

In the next chapter, we will discuss integration of NServiceBus into more modern environments and the ability to use NServiceBus in conjunction with more modern technology as the world approaches **Software as a Service (SAAS)**.

6
Using NServiceBus in the Cloud

In this chapter, we will be focusing on snippets in NServiceBus in the cloud after a very brief introduction to the cloud and some of its services. While NServiceBus has support as a service bus for the Microsoft Azure cloud, it is also a beneficial tool to integrate into other cloud technologies, as all clouds have support for third-party integration to pass data through web services.

In this chapter, we will cover:

- Introducing the cloud and NSB
- Introducing PaaS, IaaS, and SaaS
- Using Microsoft Azure
    - Introducing Azure Storage services
    - Azure Service Bus and Storage Queues
    - Azure Storage Queues and NSB
    - Azure Service Bus in NSB

Introducing the cloud and NSB

As this book is being written, NSB Version 5.0 for Azure is in the beta stage, and many updates have been made to NSB for cloud computing.

Instead of using a local SQL Server, RavenDB, or MSMQ queuing system, all the subscription data, saga persistence, timeout persistence, and queuing messages can be used in Azure Storage, as well as a message be put in the Azure Service Bus. In many cases, there is one-to-one mapping to storage tables and the data information as keys are used to retrieve data instead of SQL schemas.

NSB tools are adapting as well. ServiceInsight is growing to handle more details with sequence diagrams for debugging outside of Visual Studio and MSMQ environments. The tools are moving from being integrated locally into a physical server to being integrated into a remote server where many of the details of the server itself may be less important as servers are virtualized offsite. For instance, the future of ServiceInsight will add Saga Sequence diagrams to add more detail than the current flow diagrams. Here's what is coming soon for ServiceInsight as a sequence diagram:

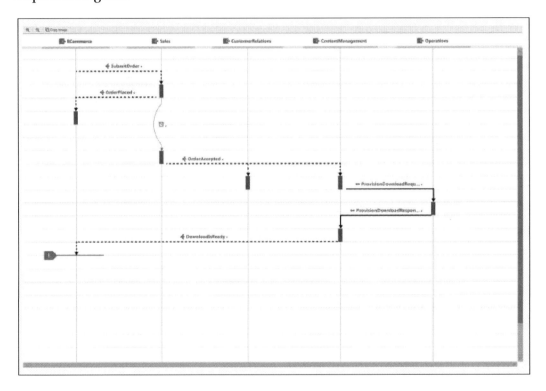

However, distributed transactions are not supported in many cloud technology queues. Other queuing platforms, such as RabbitMQ, also do not support distributed transactions. NSB Version 5.0 has the added functionality to compensate for not having distributed transactions by establishing tables that keep track of the status of transactions as they occur. This functionality allows transactional integrity where it may not exist natively.

Introducing PaaS, IaaS, and SaaS

Cloud computing services are provided using three fundamental models:

- **SaaS**: This stands for **Software as a Service**. This facilitates pay-per-use managed software services for consumption. Our responsibility is the configuration of these services that are managed by the cloud vendor. An example would be Google mail. This offers abstraction of the business software.

- **PaaS**: This stands for **Platform as a Service**. This refers to pay-as-you-go generic services for consumption. These are services such as queuing, web servers, and other services that still require business software to be built. Our responsibility includes building the business software. This offers abstraction of the services that can run business software.

- **IaaS**: This stands for **Infrastructure as a Service**. Providers of IaaS offer generic storage, networks, and servers for consumption. These are services such as storage that still require services and business software that has to be built. Our responsibility includes building the business software and creating services for the business software. This offers abstraction of the infrastructure that is running the platform services.

PaaS is a cloud computing service that provides a computing platform and a solution stack as a service. In PaaS, the cloud computing provider provides the operating systems, databases, web servers, development tools, and other services that are required to host the consumer's application.

IaaS is a level below PaaS, as it provides the virtual (as well as physical) machines, servers, storage options, load balancers, networks, and more basic components.

SaaS is a level higher than PaaS, as it is the software distribution model in which the applications themselves are hosted by a vendor or service provider and are made available to customers over a network, typically the Internet. Here is a look at how Windows Azure supports IaaS and PaaS:

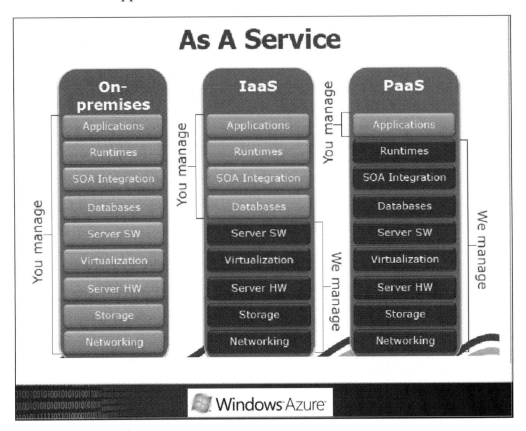

Depending on the cloud vendor, some of these terminologies may be termed slightly differently. While all cloud vendors support these components, they differ on the level of abstraction of these components. For example, Windows Azure will allow you to configure a **virtual machine (VM)** in the cloud, while others only offer SaaS models. The cloud vendor as well as licensing that are selected will dictate the limitations of your resources. In Windows Azure, you could create a free website which will require no running cost; however, your application may be shared with other websites running on that box. But if you are willing to pay more, you could have a dedicated VM with your own web server running your website. It all depends on what services you choose. In the cloud world, consumption of resources and transactions is based on licensing. There are many resources available by all cloud vendors for development in their cloud; there is also a lot of help available as their goal is to get you to utilize their cloud as much as possible, as that is their revenue stream.

Using Microsoft Azure

One of the benefits of Azure is that it can be used in Microsoft data centers around the world and developed and tested on premises before deployment using the Microsoft Azure **Software Development Kit (SDK)**. Microsoft Azure is the only cloud vendor that has an SDK to emulate all the pieces in Visual Studio 2012. The SDK allows you to even develop pieces on-site and pieces in the cloud to create hybrid solutions. The Windows Azure SDK from Microsoft is considered open source. Some important links are:

- `https://github.com/Azure`: This is the GitHub open source location
- `https://github.com/Azure/azure-sdk-tools`: This contains Microsoft Azure PowerShell tools
- `http://research.microsoft.com/en-us/projects/azure/windows-azure-for-linux-and-mac-users.pdf`: This provides Azure command-line tools for Linux and Mac operating systems
- `https://azure.microsoft.com/en-us/`: This contains *Getting Started* tutorials
- `https://manage.windowsazure.com/`: This is the Azure cloud portal
- `http://azure.microsoft.com/en-us/develop/net/samples/`: This contains Azure samples
- `http://azure.microsoft.com/en-us/gallery/store/`: This is the Azure store for add-ons
- `http://azure.microsoft.com/en-us/pricing/calculator/`: This provides the Azure cost calculator

The benefits of Microsoft Azure include the following:

- Creation of virtual machines, which include both Microsoft and Linux operating systems, for those who still wish to manage operating system functionality in IaaS.
- Ability to develop applications locally in Visual Studio using Microsoft Azure SDK tools that will work both on-site and in the cloud. This includes the ability to develop and test the functionality locally and deploy only what you wish to deploy to the cloud.
- Ability to create hybrid solutions where parts can be managed on-site and in the cloud. This comes in handy if cloud prices change and you wish to move pieces back to a premise.

- Availability of an assortment of common on-premise tools and services that work both on-site and in the cloud; for instance, using queuing and a SQL server that works locally and in the cloud.
- Development of websites quickly in the cloud through multiple tools, such as WebMatrix.
- Availability of on-demand services in using only what you need, and Microsoft tools to help report your resources and utilization for billing verification.
- Ability to move to other cloud services as your software model progresses and your needs change. Many business software applications that are developed in Azure can be used in other cloud services.
- Availability of simple data storage models in reporting and configuration of large data resources.

All of these features sum up that Microsoft allows you to develop systems on-premise or in the cloud using Microsoft tools. It does not force you to stay in the cloud or develop locally, but allows you to develop the pieces in the cloud that you see fit, and remove them from the cloud when you see fit. Many other cloud vendors may force you to use specialized tools to create solutions, which when you take them out of the cloud, will have to be completely rewritten to put into a different cloud service.

Introducing Azure Storage Services

For data management, we can create a SQL Azure database in the cloud or a SQL server on-premise, or a hybrid thereof. An easier method to keep SQL servers out of the equation is simply to call Storage Services directly, which provides rows that directly map to data. Azure Storage Services can be used both on-site through the use of the Azure SDK, and in the Azure cloud.

Azure Storage can be used to store saga data in the tables, timeout persistence in the tables, subscription information in the tables, and to transport messages in the queues.

Other than providing SQL Azure for storing data in the cloud, Azure also provides the following other types of storage, of which there are three basic types:

- **Tables**: This is based on a key-value NoSQL table format.
- **Binary Large Object (BLOB)**: This is used for binary storage, such as video files.
- **Queues**: This is used for storing messages.

- A table is a NoSQL solution instead of the relational SQL database. It can store data across multiple machines. Each table can contain partitions across multiple machines. These tables have entities with partitions and row keys to access the entity. It is a key-value store to store the data. These tables do not enforce a schema. By not using a schema, speed is enhanced and the tables are loosely coupled with the objects as they are managed by keys. The limitation is not being able to execute SQL queries, but in most cases, it can take less storage, thus less cost for the data. For tables, there is a partition key, row key, and timestamp.
- **A partition key**: This is a unique key associated with partition as a collection of all associated rows. This is used to define which partition to access; for example, the name of the table.
- **A row key**: This is a unique key to identify the row in the partition, usually a unique ID.
- **Timestamp**: This is the time at which the row is updated by Azure.

A BLOB is a group to the containers, which is just unstructured data, such as a video or audio file stored as binary storage in a data store.

The Queue storage is very similar to storing messages in MSMQ, except that the management tools are in the Azure cloud. Just as many of the Azure cloud items can be managed through the Azure SDK and Visual Studio, storage queues can also be managed through Visual Studio. We see timeout states, saga data, and subscription data in Visual Studio as shown in the following screenshot:

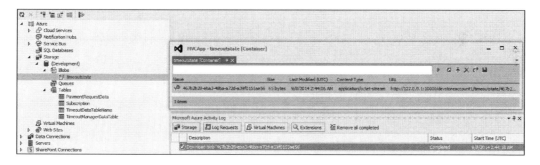

Besides using Visual Studio, there are many open source tools such as the Azure Storage Explorer found at the following:

http://azurestorageexplorer.codeplex.com/

The Azure Storage Explorer is shown in the following screenshot:

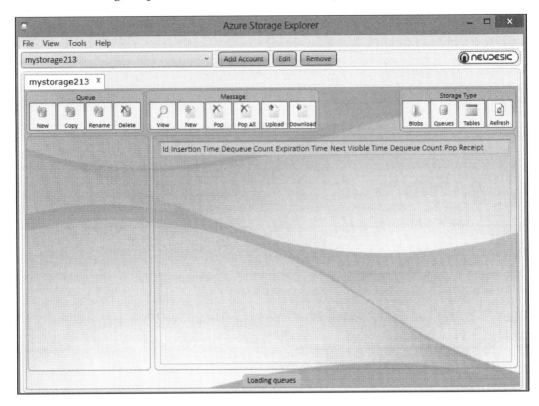

Azure Service Bus and Storage Queues

NSB supports queuing in both Azure Storage Queues and the Azure Service Bus Queues.

Please note that there are many differences between using Azure Storage Queues and Azure Service Bus Queues.

> Note the different names in this example. Service Bus Queues have more features for management such as guaranteed FIFO, while Azure Queues have less manageability built in. Visit http://msdn.microsoft.com/en-us/library/hh767287.aspx for more information.

The advantages of using Azure Storage Queues are:

- They are constructed in the storage system
- They utilize HTTP(S)
- Their capacity limit is 500 TB
- They are very cheap

The advantages of using Azure Service Bus Queues are:

- High reliability
- No limit for **time to live (TTL)**
- Support for queues, topics, and subscription
- More management tools
- TCP support for **Advanced Message Queuing Protocol (AMQP)** and **Service Bus Messaging Protocol (SBMP)**
- Certain limitations of using Azure are as follows:
    - 5 GB capacity limit
    - More expensive

Azure Storage Queues and NSB

NSB has full support for implementing Azure Queue storage on-premise or in the cloud. In this example, the solution will be in the `SagaPaymentClient - AzureSQ` directory in which we will be adding Azure Storage functionality from the previous `SagaPaymentClient` example in *Chapter 4, Saga Development*.

In our example, we will be using the Azure storage emulator to run the Storage queues locally in the Windows operating system.

We will set up the environment to run the Azure storage emulator in the following manner:

1. The emulator can be installed through a standalone package found at http://www.microsoft.com/en-us/download/details.aspx?id=42317 or from the Microsoft Azure SDK. The Azure SDK can be installed by using the Web Platform installer found at http://www.microsoft.com/web/downloads/platform.aspx.

The following screenshot displays the **Web Platform Installer 5.0** window:

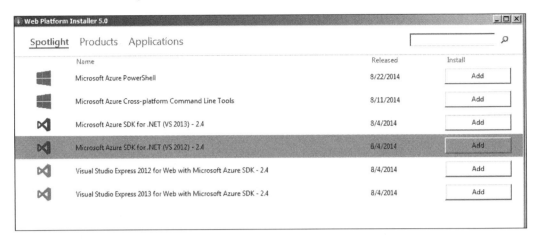

2. Start the Azure storage emulator from the tool that was installed: Windows Azure Storage Emulator, as shown in the following screenshot:

We will set up the NSB to run Azure Storage Queues in the following manner:

1. Install the reference from either NuGet or the package manager for the project for NServiceBus using `PM> Install-Package NServiceBus.Azure.Transports.WindowsAzureStorageQueues` and install the `NServiceBus.Azure` reference `PM> Install-Package NServiceBus.Azure`.

2. Ensure that you have the correct versions of the Windows Azure Storage by using `Install-Package WindowsAzure.Storage`.

3. Set up the configuration on the IBus configurations. In the `MySaga` project, we will use Azure for the saga, the queue, the timeout, and the subscription as follows:

```
Configure.With()
    .DefaultBuilder()  // Autofac Default Container
    .UseTransport<NServiceBus.AzureStorageQueue>()  // Azure Storage Queues
    .AzureSubscriptionStorage()
    .AzureSagaPersister()
    .UseAzureTimeoutPersister()
    .UnicastBus(); // Create the default unicast Bus
```

4. We will then set the `App.config` settings which include the `AzureSagaPersisterConfig`, `http://AzureTimeoutPersisterConfig`, `AzureSubscriptionStorageConfig`, and `NServiceBus/Transport` as we set it to `UseDevelopmentStorage=true` to use the Storage Emulator. The code is shown in the following screenshot:

```
<section name="AuditConfig" type="NServiceBus.Config.AuditConfig, NServiceBus.Core" />
<section name="AzureProfileConfig" type="NServiceBus.Config.AzureProfileConfig, NServiceBus.Hosting.Azure" />
<section name="AzureSubscriptionStorageConfig" type="NServiceBus.Config.AzureSubscriptionStorageConfig, NServiceBus.Azure" />
<section name="AzureSagaPersisterConfig" type="NServiceBus.Config.AzureSagaPersisterConfig, NserviceBus.Azure" />
<section name="AzureTimeoutPersisterConfig" type="NServiceBus.Config.AzureTimeoutPersisterConfig, NserviceBus.Azure" />
</configSections>
<!-- Azure Settings-->
<AzureSagaPersisterConfig ConnectionString="UseDevelopmentStorage=true" />
<AzureTimeoutPersisterConfig ConnectionString="UseDevelopmentStorage=true" />
<AzureSubscriptionStorageConfig ConnectionString="UseDevelopmentStorage=true" />
<connectionStrings>
    <add name="NServiceBus/Transport" connectionString="UseDevelopmentStorage=true" />
```

We will have to repeat each of these steps for each project as needed.

We have set up the functionality to use the Azure storage queues. When running the application, you should have the solution in the `Payment_WCFService` directory running as this is the client with the saga to send messages to the WCF service.

Using NServiceBus in the Cloud

When we run the client, we will get the MVCApp website to send a message to the WCF client through a saga to the WCF server. It will appear as follows:

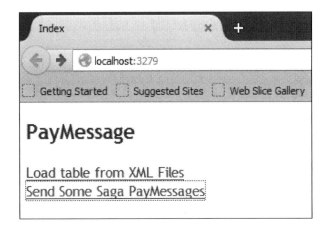

We will also have the `MySaga` and WCF client running as well to create queues and persistence in the Azure emulator. In Visual Studio 2012, by going to **View** | **Server Explorer** | **Azure** | **Storage**, we can view the queues and messages as they are running, as follows:

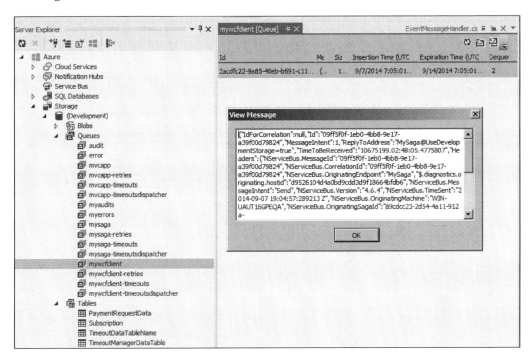

We can also examine the tables as well as the persistent saga data as follows:

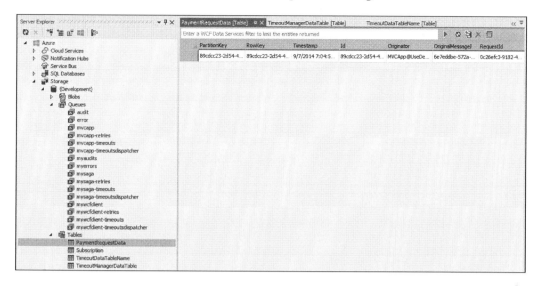

So we have Azure storage queues running in the emulator in a local development using NServiceBus with very little effort. Again, NSB handles most of the work as we focus on the business logic and configurations.

Azure Service Bus in NSB

The Windows Azure Service Bus provides a hosted, secure, and widely available infrastructure for widespread communication between different messaging endpoints to include web services. The Service Bus communicates via three methods:

- **Queues**: These deal with one-to-one messaging through queues
- **Topics**: These deal with one-to-many publish-subscribe messages from one publishing endpoint to many subscriber endpoints
- **Relays**: These deal with one-to-one relationships between requests and replies that talk directly to the endpoints

Using NServiceBus in the Cloud

The Azure Service Bus can be created in the Azure portal by first creating the Service Bus namespace as shown in the following screenshot:

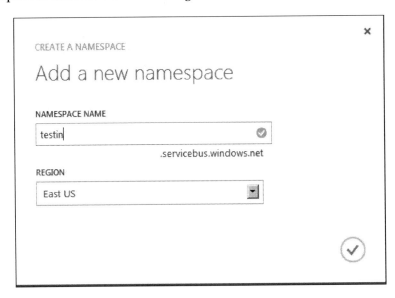

We can create the Service Bus Queues, Topics, and Relays in the Azure portal, and we can also manage the Service Bus through Visual Studio after it is initially created in the Azure portal. The Service Bus will use a primary key and connection string to be accessed, not too dissimilar to a connection string to SQL database.

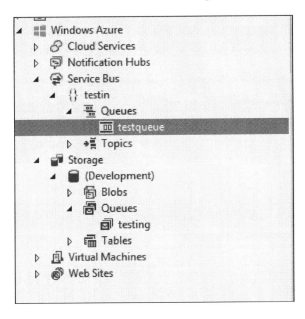

Chapter 6

When creating an Azure Service Bus queue, we can see the additional configuration settings that are offered in this creation and not offered in Storage queues as shown in the following screenshot:

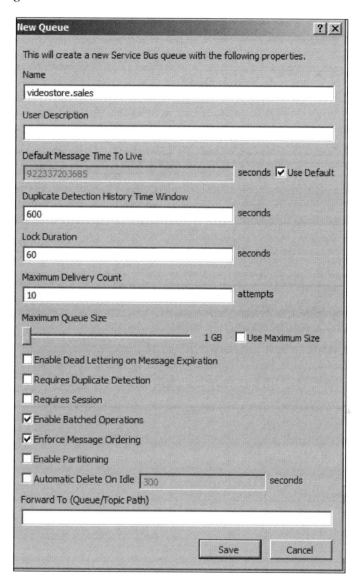

Using NServiceBus in the Cloud

There are additional tools for exploring the Azure Service Bus, such as the Server Bus Explorer found at `http://code.msdn.microsoft.com/windowsazure/Service-Bus-Explorer-f2abca5a`.

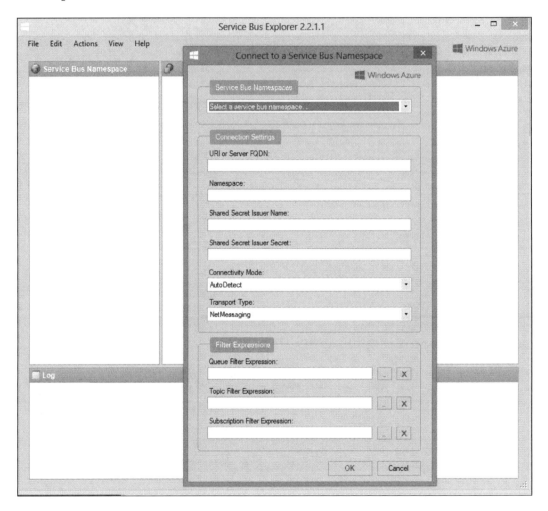

We will start with our example, which can either be developed on-site or in the cloud. We will do this one in the cloud. We will use the previous example `SagaPaymentClient` and have a new directory with the Service Bus example called `SagaPaymentClient - AzureServiceBus`. The example will work in a similar manner, except that we will be sending the queues across the Azure Service Bus in the cloud. We will be testing our queuing namespace in the cloud.

For examples, we can also use the NServiceBus example for the Azure Cloud Service Bus queues from:

`https://github.com/Particular/NServiceBus.Azure.Samples/tree/master/VideoStore.AzureServiceBus.Cloud`

We will set up the NSB to run Azure storage queues in the following manner:

1. Install the reference from either NuGet or the package manager for the project for NServiceBus using `PM> Install-Package NServiceBus.Azure.Transports.WindowsAzureServiceBus`.

    You may need to match the compatible versions of NServiceBus core versions with the corresponding NServiceBus Azure versions. One thought is to uninstall the current NServiceBus references, and let the NServiceBus Azure references install the required NServiceBus core references.

2. Set up the configuration on the IBus configurations. In the `MySaga` project and `EndpointConfig.cs` file, we use the AzureServiceBus for the queue as follows:

```
Configure.With()
    .DefaultBuilder()  // Autofac Default Container
    .UseTransport<NServiceBus.AzureServiceBus>()  // Azure Storage Queues
    .AzureSubscriptionStorage()
    .AzureSagaPersister()
    .UseAzureTimeoutPersister()
    .UnicastBus(); // Create the default unicast Bus
```

Using NServiceBus in the Cloud

3. We will then set the `App.config` or `Web.config` file depending on the type of project. We need to get the Service Bus connection string settings.

   We can retrieve the Azure Service Bus connection string from the tab shown in the following screenshot by clicking on the **CONNECTION INFORMATION** as shown below:

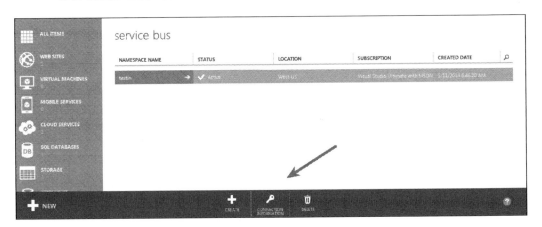

We copy and then store the connection information in the app or web configuration file depending on what kind of application we are building into `NserviceBus/Transport` and `Microsoft.ServiceBus.ConnectionString` as we see in the following:

```xml
<section name="AzureSubscriptionStorageConfig" type="NServiceBus.Config.AzureSubscriptionStorageConfig, NServiceBus.Azure" />
<section name="AzureSagaPersisterConfig" type="NServiceBus.Config.AzureSagaPersisterConfig, NserviceBus.Azure" />
<section name="AzureTimeoutPersisterConfig" type="NServiceBus.Config.AzureTimeoutPersisterConfig, NserviceBus.Azure" />

</configSections>

<AzureSubscriptionStorageConfig ConnectionString="UseDevelopmentStorage=true" />
<AzureSagaPersisterConfig ConnectionString="UseDevelopmentStorage=true" />
<AzureTimeoutPersisterConfig ConnectionString="UseDevelopmentStorage=true" />

<!-- NHibernate Settings-->
<connectionStrings>
    <add name="NServiceBus/Transport" connectionString="Endpoint=sb://testin.servicebus.windows.net/;SharedSecretIssuer=owner;SharedSecretValue=NB64RL1zcZPdx3EORsFI6lS6i8DUyWY7eh2fjIpmf44=" />
</connectionStrings>
<!-- specify the other needed NHibernate settings like below in appSettings:-->
<appSettings>
    <!-- Service Bus specific app setings for messaging connections -->
    <add key="Microsoft.ServiceBus.ConnectionString" value="Endpoint=sb://testin.servicebus.windows.net/;SharedSecretIssuer=owner;SharedSecretValue=NB64RL1zcZPdx3EORsFI6lS6i8DUyWY7eh2fjIpmf44=" />
</appSettings>
```

> We also set the `AzureSagaPersisterConfig`, `AzureTimeoutPersisterConfig`, `AzureSubscriptionStorageConfig` configurations to still use the Azure Storage for the database settings. We added the `AzureSubscriptionStorageConfig` configuration to all the projects.

We will have to repair the settings to the projects as needed to set the transport to `AzureServiceBus`.

We now have the configuration changes moved to use `AzureServiceBus`. When we run the application we can view all the queues from the Azure management portal as follows:

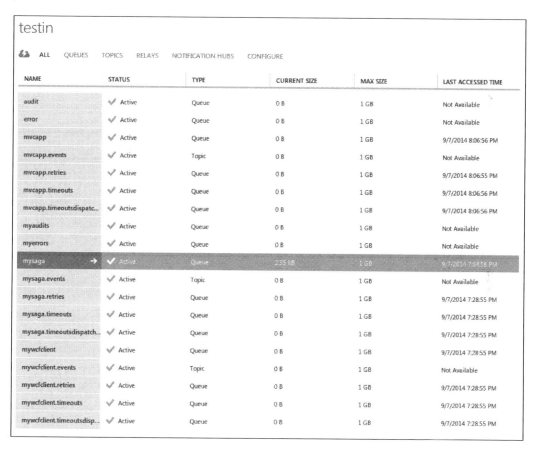

Using NServiceBus in the Cloud

We can see that `MySaga` has a message that we sent it from the MVCApp client as follows:

Instead of using the Azure Management Portal, we can also view the messages from the Visual Studio Server Explorer as we add a connection to match the connection string to the queue:

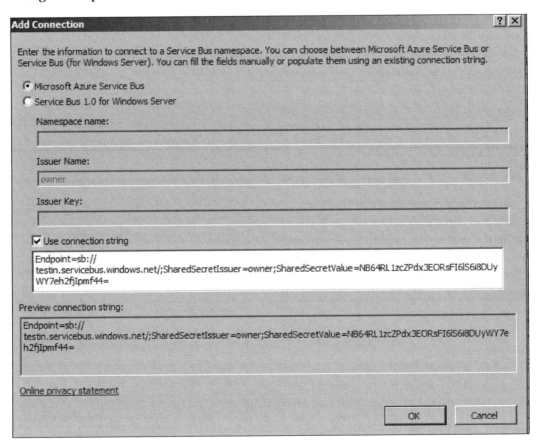

We view the connections in the Visual Studio Server Explorer:

> There are topics, the publish-subscribe messaging, and queues that are one-to-one messages.

We now have a running Azure Service Bus in NSB.

Summary

In this chapter, we took a deeper dive into **Software as a Service (SaaS)** and how NServiceBus ties into cloud computing. We had a very brief introduction to the cloud and some of its services. We discussed the Azure Storage Services and Azure Service Bus as they apply to providing persistence and message queues. Visual Studio has many tools for building applications both on-premise and in the Azure cloud. We walked through a sample in each, where we stored all the timeouts, saga, and subscription information in Azure.

Index

A

ActiveMQ
 about 139, 168
 source code 139-142, 168-171
 using, in NSB 141
Advanced Encryption Standard (AES)
 URL 30
Apache Active Message Queue. *See* **ActiveMQ**
ASP.NET MVC
 about 94
 overview 94, 95
Azure cloud portal
 URL 177
Azure command-line tools
 URL 177
Azure cost calculator
 URL 177
Azure samples
 URL 177
Azure SDK
 Relays 181
Azure Service Bus
 in NSB 185-193
 Queues 185
 Relays 185
 Topics 185
Azure Service Bus Queues
 advantages 181
Azure storage emulator
 Relays 181
Azure Storage Explorer
 URL 179
Azure Storage Queues
 about 180
 advantages 181
 and NSB 181-185
Azure Storage Services
 about 178, 179
 Binary Large Object (BLOB) 178
 partition key 179
 queues 178
 row key 179
 tables 178
 timestamp 179
Azure store
 for add-ons, URL 177

B

Binary Large Object (BLOB) 178

C

cloud
 and NSB 173, 174
CustomChecks
 for ServicePulse 88-91

D

DataBus pattern 21, 22
deployment 55, 56

E

e-mail saga notification, sample
 about 146-148
 saga project 150-154
 solution, testing 155-157
 XAML, using 148-150
Enterprise Service Bus (ESB) 10-12

ESB 10-12
ESB designs
  practical needs 13
event-driven jobs 14
Extensible Application Markup
     Language (XAML)
  URL 146
Extensible Application Markup Language
     (XAML), using
  for e-mail saga notification 148-150
  for SFTP saga 160

G

gateway pattern
  about 18
  source code 19

I

Infrastructure as a Service (IAAS) 175

J

Java Message Service (JMS) 139, 168
Java Runtime Environment (JRE) 139
JavaScript Object Notation (JSON) 62

L

Language Integrated Query (LINQ) 94

M

message
  flow 48-55
message encryption patterns
  about 30
  source code 30, 31
message mutation patterns
  about 28
  source code 29
Microsoft Azure
  about 177, 178
  features 177, 178
Microsoft Azure PowerShell tools
  URL 177
Microsoft Entity Framework (EF) 94
Microsoft Message Queuing (MSMQ) 12

Microsoft Task Scheduler 22
Model-View-Controller (MVC) 94

N

NSB
  about 9
  and Azure Storage Queues 181-185
  and cloud 173, 174
  Azure Service Bus 185-193
  benefits 12
  components 59
  host, URL 71
NSB hosting
  URL 56
NSB version 4
  to 5, upgradation 39
  to 5, upgradation from 39, 41
NSB version 5
  features 39-41
NServiceBus. *See* NSB
NServiceBus, adding to MVC
  about 120-122
  integration tests, performing 130-134
  message handler unit testing 123-125
  saga handler unit testing 128-130
NServiceBus.RabbitMQ package 138

O

object-relationship mapper (ORM) 95
  boilerplate code, generating 95
  development, speeding 95
  OO supporting 95

P

PaaS 175
patterns
  DataBus pattern 21, 22
  gateway pattern 18
  message encryption patterns 30
  message mutation patterns 28
  publish-subscribe pattern 15, 16
  request-reply pattern 17
  saga design pattern 32
  ScaleOut pattern 32
  timeout patterns 22-28

Platform as a Service. *See* PaaS
publish-subscribe pattern 15, 16

R

RabbitMQ
  about 135
  administrating 135
  endpoints, changing 137, 138
  source code 137
  URL, for development tools 135
  URL, for documentation 135
  URL, for NServiceBus RabbitMQ hands-on lab 135
  URL, for NServiceBus samples 135
  URL, for NSeviceBus.RabbitMQ source code 135
  URL, for Windows installation 135
request-reply pattern 17
requisites, NSB WCF Integration
  configuration for NSB with WCF 98
  message handler 98
  request message structure 98
  response message structure 98
  web service 98

S

SaaS 175, 176
saga
  advantages 34-36
  deployment 165, 166
  e-mail saga notification, sample 146-148
  SFTP saga, sample 158, 159
  source code 146
  through ServiceMatrix 84-88
  workflow 41-47
saga design pattern
  about 32
  ConfigureHowToFindSaga() 34
  IAmStartedByMessages<IMessage> 32
  IHandleMessages<IMessage> 34
  Saga<IContainSagaData> 33
saga development
  web services 96
ScaleOut pattern 32
Secure File Transfer Protocol. *See* SFTP

Service Broker 10
ServiceControl
  about 62, 63
  plugins, adding 65, 66
  ServiceControl.Plugin.CustomChecks 65
  ServiceControl.Plugin.DebugSession 64
  ServiceControl.Plugin.Heartbeat 65
  ServiceControl.Plugin.SagaAudit 65
  URL 60, 62
ServiceControl plugins
  for CustomChecks 77
  for DebugSessions 77
  for heartbeats 77
  for SagaAudits 77
ServiceInsight
  about 56-69
  endpoints 71
  infrastructure 71
  libraries 71
  services 71
  solution, creating 69-74
  URL 60
ServiceMatrix
  about 59, 60
  sagas through 84-88
  URL 68
service-oriented architecture (SOA) 10, 96
ServicePulse
  about 60, 61
  CustomChecks for 88-91
SFTP
  about 12, 145
  URL 158
SFTP saga, sample
  about 158, 159
  features 159
  messaging process, changing 160, 161
  SFTP test environment, setting up 161-165
  XAML, using 160
Simple Object Access Protocol (SOAP) binding 96
SOA patterns 15
Software as a Service. *See* SaaS
Software Development Kit (SDK) 177
SQLNinja
  URL 94

T

timeout patterns 22-28

V

virtual machine (VM) 176

W

WCF
  about 96
  source code 96
WCF client
  creating 111
  design, revisiting 116-118
  service reference, adding 112-114
  service reference, calling 114-116
  source code 118, 119

WCF server
  configuration, adding 103-105
  considerations, for deployment 110
  creating 97-99
  message handler, adding 102
  messages, adding 100, 101
  tracing, adding 106-109
  web service, viewing 110
Web Service Definition Language (WSDL) 10
Web Services Description Languages (WSDL) 96
Windows Communication Foundation (WCF) 13
Windows Presentation Framework (WPF) 145, 146
Windows Task Scheduler 14

Thank you for buying
Learning NServiceBus Sagas

About Packt Publishing

Packt, pronounced 'packed', published its first book *Mastering phpMyAdmin for Effective MySQL Management* in April 2004 and subsequently continued to specialize in publishing highly focused books on specific technologies and solutions.

Our books and publications share the experiences of your fellow IT professionals in adapting and customizing today's systems, applications, and frameworks. Our solution based books give you the knowledge and power to customize the software and technologies you're using to get the job done. Packt books are more specific and less general than the IT books you have seen in the past. Our unique business model allows us to bring you more focused information, giving you more of what you need to know, and less of what you don't.

Packt is a modern, yet unique publishing company, which focuses on producing quality, cutting-edge books for communities of developers, administrators, and newbies alike. For more information, please visit our website: www.packtpub.com.

About Packt Enterprise

In 2010, Packt launched two new brands, Packt Enterprise and Packt Open Source, in order to continue its focus on specialization. This book is part of the Packt Enterprise brand, home to books published on enterprise software – software created by major vendors, including (but not limited to) IBM, Microsoft and Oracle, often for use in other corporations. Its titles will offer information relevant to a range of users of this software, including administrators, developers, architects, and end users.

Writing for Packt

We welcome all inquiries from people who are interested in authoring. Book proposals should be sent to author@packtpub.com. If your book idea is still at an early stage and you would like to discuss it first before writing a formal book proposal, contact us; one of our commissioning editors will get in touch with you.

We're not just looking for published authors; if you have strong technical skills but no writing experience, our experienced editors can help you develop a writing career, or simply get some additional reward for your expertise.

Learning NServiceBus

ISBN: 978-1-78216-634-4　　　Paperback: 136 pages

Build reliable and scalable distributed software systems using the industry leading .NET Enterprise Service Bus

1. Replace batch jobs with a reliable process.
2. Create applications that compensate for system failures.
3. Build message-driven systems.

Multithreading in C# 5.0 Cookbook

ISBN: 978-1-84969-764-4　　　Paperback: 268 pages

Over 70 recipes to help you learn asynchronous and parallel programming with C# 5.0 quickly and efficiently

1. Delve deep into the .NET threading infrastructure and use Task Parallel Library for asynchronous programming.
2. Scale out your server applications effectively.
3. Master C# 5.0 asynchronous operations language support.

Please check www.PacktPub.com for information on our titles

CryENGINE Game Programming with C++, C#, and Lua

ISBN: 978-1-84969-590-9   Paperback: 276 pages

Get to grips with the essential tools for developing games with the awesome and powerful CryENGINE

1. Dive into the various CryENGINE subsystems to quickly learn how to master the engine.

2. Create your very own game using C++, C#, or Lua in CryENGINE.

3. Understand the structure and design of the engine.

Enterprise Integration with WSO2 ESB

ISBN: 978-1-78328-019-3   Paperback: 92 pages

Over 15 recipes to calibrate seamless modularity to SOA and address commonly-faced enterprise integration challenges with a zero-code approach

1. Learn how to implement the mostly-used Enterprise Integration Patterns with WSO2 ESB.

2. Discover how to integrate WSO2 ESB with FIX, HL7, and SAP gateways.

3. Understand the key concepts behind WSO2 ESB, and find optimized recommendations for deploying WSO2 ESB in a production setup.

Please check **www.PacktPub.com** for information on our titles

Made in the USA
Middletown, DE
13 April 2018